Old Dominion, Industrial Commonwealth

Studies in Early American Economy and Society from the
Library Company of Philadelphia

Cathy Matson, Series Editor

Old Dominion, Industrial Commonwealth

Coal, Politics, and Economy in Antebellum America

Sean Patrick Adams

The Johns Hopkins University Press

Baltimore

© 2004 The Johns Hopkins University Press
All rights reserved. Published 2004
Printed in the United States of America on acid-free paper

Johns Hopkins Paperback edition, 2009
9 8 7 6 5 4 3 2 1

The Johns Hopkins University Press
2715 North Charles Street
Baltimore, Maryland 21218-4363
www.press.jhu.edu

The Library of Congress has catalogued the hardcover edition of this book as follows:
Adams, Sean P.

Old Dominion, industrial commonwealth : coal, politics, and economy in antebellum
America / Sean Patrick Adams.

p. cm. — (Studies in early American economy and society from the Library
Company of Philadelphia)

Includes bibliographical references and index.

ISBN 8-8018-7968-X (hardcover : alk. paper)

1. Coal trade—Pennsylvania—History. 2. Coal trade—Virginia—History. I. Title.
II. Series.

HD9554.U63P43 2004

338.2′724′0975409034—dc22

2004001685

A catalog record for this book is available from the British Library.

ISBN 13: 978-0-8018-9400-8
ISBN 10: 0-8018-9400-X

Contents

Figures, Maps, and Tables

Series Editor's Foreword

$\mathcal{T}$his volume in the Studies in Early American Economy and Society series continues the collaborative effort between the Johns Hopkins University Press and the Library Company of Philadelphia's Program in Early American Economy and Society (PEAES). Since its inception in 2000, PEAES has undertaken as part of its central mission the advancement of research on the early American economy through the antebellum era as well as the presentation of some of the important fruits of that research to a wide readership. Sean Adams' present study of the comparative political economies of coal in Pennsylvania and Virginia represents a signal contribution to the dialogue about early American economic relationships which such research engenders.

Scholars active in researching, writing, and teaching sometimes note our glaring collective failure to create conversations across the professional disciplines, and even within them, about the early American economy. Clearly, bridging the gulf between historical inquiry and economic issues offers great possibilities for enriching our understanding about how early Americans constituted their lives before industrialization. In this volume, Adams' case study of regional differences in thinking about legislating policies toward mining, moving, and consuming coal brings together the state-level institutional development of early national Pennsylvania and Virginia with the regional divergence of economic cultures in the North and the South. Comparing the two states' contentious public discourses on internal improvements, factious policy initiatives related to coal, and economic cultures of slavery and entrepreneurship, Adams makes important arguments about the active agency of state governments in responding to the abundance of a mineral fuel which became essential to American industrialization. His broad source base and array of voices lead Adams away from portraits of an emerging "national identity" and the oversimplified views of sectional divergence present since the 1940s in our economic histories of the early republic and toward a multidisciplinary, nuanced argument about regional economic political economy.

Cathy Matson, University of Delaware, and
Director, Program in Early American Economy and Society,
The Library Company of Pennsylvania

Acknowledgments

Over the past decade of working on this project, I've learned a great deal about coal but also rediscovered quite a bit about the coal industry's place in American life. I grew up in Morgantown, West Virginia, which in the 1980s was a university town still surrounded by an active coal industry, but I never thought critically about how the local mines affected my community. I had no personal or familial connections to the trade, but the rhythm of coal mining was there. On wintry mornings radio announcers would wind their way through the various mine closings before getting to the matter of which schools—a matter that interested me much more—would also be closed as a result of the weather. Outside of town a fleet of coal trucks dusted buildings with fine soot and forced young drivers like myself to develop quick reflexes on narrow country roads. On a more somber note the raspy cough or pronounced limp often seen around Courthouse Square reminded folks of the coal industry's high human cost. Worst of all, the word of a fire, a collapse, or flooding of a nearby mine spread quickly in Morgantown and permeated even the most isolated pockets of conversation. Although I was steeped in this environment, I never planned to write about coal. This book is the result more of my academic interests than personal experiences—I chose coal as a case study to explore larger questions of regional development in the nineteenth century and the nature of the Industrial Revolution in the United States. But I found, as many scholars do, that my research in a roundabout way became related to my own memories and history.

Yet all the recovered memories in the world won't allow you to complete such a project without the help of friends, colleagues, and kindhearted professional historians, archivists, and librarians—I cannot list them all. This project began at the University of Wisconsin–Madison, under the watchful eye of Colleen Dunlavy. From my initial contact with her in a first-year research seminar on comparative industrialization, her sharp analysis of historical problems, generosity, and infinite patience with my writing helped me every step of the way. She served as a wonderful advisor and model scholar. Diane Lindstrom also deserves special mention (and none of the blame) for the final state of this project, for she taught me a great deal about economic history and the profession in general. At Madison the ideas flowed freely among a number of talented graduate students, and I was lucky to gain the insights of Steve Burg, Tim Cleary, Ted Franz, Dan Graff, Charlotte Haller, Tony Harkins, Steve Kolman, Jon Pollack, Ty Priest, Jon Rees, Bethel Saler, and Andrew Schrank. My research benefited greatly from presentations to both the members of the Industrial History Reading Group and of John Cooper and Tom Archdeacon's Mellon Summer Seminar, most notably Thomas Andrews Tracey Deutsch, Eric Morser, Lisa Tetrault, Tim Thering, and Susie Wirka. Final thanks should go to the heroic staff of the State Historical Society of Wisconsin's library and archives divisions. I haunted the government documents stacks of the SHSW for years, and they never kicked me out. This project, like so many others coming out of the UW program, would have been impossible without the patience and professionalism of Jim Danky, Michael Edmonds, and a host of other marvelous librarians.

A number of individuals and organizations helped with the research away from Madison. Roger Horowitz, Glenn Porter, and Philip Scranton at the Hagley Library; Linda Shopes at the Pennsylvania Historical and Museum Commission; Nelson Lankford and Frances Pollard at the Virginia Historical Society; John Van Horne, James Green, and Phillip Lapsansky of the Library Company of Philadelphia; Wendy Woloson of the Program in Early American Economy and Society; and Roy Goodman and Robert Cox of the American Philosophical Society all helped immeasurably with this project, and their institutions generously provided grants of time and money. The library and archival staffs at the Library of Virginia, the Pennsylvania Historical Society, the Alderman Library at the University of Virginia, and the West Virginia Collection in Morgantown, West Virginia, were also invaluable to this study. I also want to acknowledge the American Historical Association, the Economic History Association, the Newcomen Society of the United States, and the College of Arts and Sciences at the University of Central Florida for financial support during the research and writing of this book.

I have been blessed to be part of a wide community of scholars who have been generous with their time and criticism. The following scholars offered commentary on this project in various forms: Stefan Berger, John Bezís-Selfa, Susanna Delfino, David Koistinen, Ken Lipartito, John Majewski, Scott Nelson, Margaret Newell, Peter Onuf, Donna Rilling, Andy Schocket, Robert Wright, and the anonymous readers for the *Virginia Magazine of History and Biography*. A National Historical Records and Publications Commission Fellowship at the Frederick Douglass Papers when they were at West Virginia University proved critical for my professional development. Jack McKivigan, the project's editor, read portions of the manuscript and has been a supportive presence. Diane Barnes talked to me about coal (always a plus) and shared some insights from her own important work on Virginia artisans. Ron Lewis of the WVU Department of History was kind enough to let me sit in on his Appalachian History seminar, which helped inform my chapter on West Virginia. A number of colleagues at the University of Central Florida have also read and improved portions of this work. Special thanks go to Carole Adams, Rosalind Beiler, Spencer Downing, and Craig Friend. Earlier portions of this research have appeared in *Essays in Economic and Business History*, the *Virginia Magazine of History and Biography*, and *Business and Economic History*, and I thank the editors of each journal for their permission to use some of the material in the present volume.

In the later stages of this project I encountered a number of readers and editors who have helped immensely in the struggle to shape a book and helped to navigate the complex worlds of academic publishing. Richard John has provided encouragement and focus to my work on coal for some time. Not only did he comment on an early version of the work, but he also read the entire manuscript with a sharp analytical eye. Cathy Matson took an active role in developing this project, as director of the Program in Early American Economy and Society and as a critical reader. She has served dual duties as both a critical editor of the manuscript and coordinator of its publication. I would also like to thank John Larson, who twice read this manuscript for the Johns Hopkins University Press. His comments provided direction and focus to this project; my only hope is that I followed them well. And last, but not least, I want to thank Robert J. Brugger and Melody Herr of the Johns Hopkins University Press for their professional and patient interactions with a rookie author. Bill Nelson created the wonderful maps in the pages that follow, and Elizabeth Gratch provided expert copyediting. All of their efforts have made this a better book—any remaining deficiencies fall squarely upon my shoulders.

I should close these acknowledgments with a mention of family, because I couldn't have completed this book without them. My father, Donald Adams, led me to develop a love of history at an early age and helped me negotiate the twists and turns of the academic profession. My mother, Joyce Adams, has continued to encourage my scholarly pursuits at the same time that she developed her own important career in social work. Elizabeth Adams, my sister, is an educator and an inspiration in her own right. My final thanks go to the most important person in my life, Juliana Barr. Through difficult circumstances she has helped me in ways that I cannot measure. Through perceptive readings of my temperament Juliana knows when to cajole, when to encourage, and when to ignore. Without her help I couldn't have finished this project. She is also one of the best scholars that I've known, and I am one of the luckiest for her presence in my life.

Old Dominion, Industrial Commonwealth

Introduction

The Political Economy of Coal

*I*n the spring of 1796 Benjamin Henry Latrobe, the famed engineer and architect, toured the coalfields outside of Richmond, Virginia. A visit to the mines on the south side of the James River formed a deep impression upon the thirty-one-year-old emigrant from Great Britain. "Such a mine of Wealth, exists I believe nowhere else! A Rock, a *Mountain* of Coal sunk down 30 feet from the Surface, bored 10 feet more, and yet no substratum found!" he wrote on 19 April. "The *open* pit is about 50 Yards square and about 30 feet deep. Many Works or Drifts are from thence carried into the body of Coal, 5 feet wide. More than half the Coal still remains in Pillars supporting the roof. I hope to have another and better opportunity of examining this wonderful Mass of Coal, till then I postpone further description." Later in the summer Latrobe wrote from the Petersburg area, "From every indication I have no doubt of the country abounding, *this far,* in Coal." Latrobe, knowing the value of coal to the industrial economy of the United Kingdom, drew upon his familiarity with mineral deposits to describe the vast potential of the Old Dominion. "I have been fortunate in my conjecture," he wrote on 17 June "founded upon the analogy of this part of Virginia and many parts of Europe with which I am acquainted I hope I shall be so as to the vehicle of Coal."[1]

According to other observers, too, the Old Dominion seemed destined to serve as the center of America's coal trade. As the former colonies struggled to establish a stable economy after independence, they would need increasing amounts of mineral fuel from domestic sources. The mines of the Richmond basin appeared ready to answer this need without any major competition. "The owners of the coal mines of Virginia," Pennsylvania's Tench Coxe noted in 1794, "enjoy the monopoly of all the supplies for the manufacturers of the more northern states, who live in the sea ports; a demand which is increasing rapidly." To the patriot, the manufacturer, and the political economist of the early republic, the coalfields of the Old Dominion held enormous promise. By 1808 Latrobe's admiration for the Richmond-area mines had not waned. In a contribution to Albert Gallatin's famous report on internal improvements, he argued that "upon the coal mines of James river our Atlantic sea ports will soon become dependent for their chief supply of fuel." "That dependence exists already in respect to the fuel required for a variety of manufactures," he continued, "and even now the smiths within 10 miles of our sea ports, require in order to carry on advantageous business, a supply of Virginia coal."[2]

To the north of the Richmond basin, however, Virginia coal had a potential rival. Yet the landowners there, in the mountains of eastern Pennsylvania, seemed less optimistic about their own massive reserves of anthracite coal. Pennsylvania anthracite, unlike Richmond's bituminous coal, had a less than stellar reputation as both a domestic and industrial fuel. Although some local blacksmiths adopted anthracite, it was not widely accepted in urban markets. Even in Philadelphia, conventional wisdom held that anthracite was difficult, if not impossible, to light in fireplaces, stoves, and furnaces. In 1803 Philadelphians spread nearly thirty tons of Lehigh anthracite on their sidewalks in place of gravel; after watching it smother a fire in a trial run, they had no better use for it. A decade later George Shoemaker took nine wagonloads of anthracite to Philadelphia and gave seven of them away. "The result was against the coal," one observer recalled, "those who tried them, pronounced them stone and not coal, good for nothing, and Shoemaker an imposter." Even the city's waterworks, whose two steam engines for pumping Philadelphia's water supply burned both wood and coal, shunned Pennsylvania's "stone coal" and burned over a ton of Richmond bituminous every day by 1809.[3]

Virginia, not Pennsylvania, thus stood poised to serve as the young nation's provider of mineral fuel. Small-scale operators in the Richmond Basin shipped their jet-black mineral all over the eastern seaboard and enjoyed the "first mover" advantage so critical to new industrial ventures. Preliminary reports from Virginia's wild, unsettled western counties, moreover, described massive seams of bituminous coal that lay close to the surface. Although the mines in this area saw limited development during the early republic, the mineral assets of the Ohio and Kanawha valleys seemed boundless and easily accessible. On either side of the Appalachians the Old Dominion was truly blessed with mineral wealth.

Richmond's leadership in the American coal industry, however, quickly unraveled, and Pennsylvania's combination of anthracite in its eastern mountains and bituminous coal in its western counties proved to make a larger impact upon nineteenth-century economic growth. By 1860 Pennsylvanians mined over fifteen million tons of coal, or 78 percent of the nation's total, while the Old Dominion's production languished at just under a half-million tons, or 2.4 percent of the national amount. By 1875 Pennsylvania's production had topped thirty-five million tons, whereas Virginia and West Virginia's combined total was 1.5 million, or less than 3 percent of U.S. coal production. Pennsylvania's dominance in coal, moreover, aided the state in developing a vibrant industrial economy by the mid-nineteenth century. The state's iron and steel industry blossomed in response to the nearby presence of anthracite and bituminous mines, and the smoggy cityscapes of Pittsburgh and Philadelphia testified to the value of cheap coal to urban growth and manufacturing output. Pennsylvania, it seems, fulfilled the destiny that Benjamin Latrobe and others had outlined for Virginia. No observers of the early nineteenth century could have predicted the rapid decline of the Old Dominion's coal industry during the antebellum decades or the simultaneous rise of Pennsylvania as the nation's premier supplier of mineral fuel.

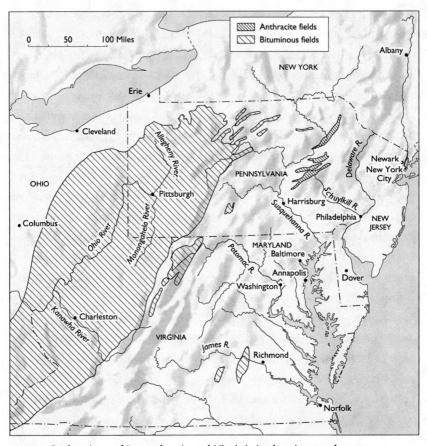

Map 1. Coal regions of Pennsylvania and Virginia in the nineteenth century

The story of Virginia and Pennsylvania coal embodies a key problem in the history of American industrialization: the question of regional divergence. How could two states with similar resource endowments embark upon radically different paths? Why did coal lay dormant in the Old Dominion throughout most of the antebellum period while Pennsylvanians raised millions of tons at the same time? Over the years scholars have attributed the growing rift between northern and southern industrial development to various and sundry causes; they have targeted divergent free and slave labor systems, the different natural resources prevalent in each region, and the North's more dynamic urban development. The major culprit for southern "arrested indus-trialization" is almost always slavery.[4]

Yet rarely do scholars examine the political economies of northern and southern states in a comparative perspective, as state governments are often consigned to a subordinate status among the causal factors for differences in northern and southern developmental paths. In most accounts political insti-tutions merely reflect economic or social trends or serve as convenient labels for regions; they rarely take on an active role in regional economic divergence. After all, slavery in the South predated the political birth of the United States by a century and a half. But, then, why did northern and southern states en-joy varying levels of industrial growth? Could this divergence be fundamen-tally linked to political factors—namely, to the role of state governments that created an institutional environment for dynamic or lethargic development?

This study compares the development of the coal industry in Virginia and Pennsylvania to contrast the ways that each state developed its mineral resources from the early nineteenth century through the emergence of in-dustrial markets and the concurrent rise of laissez-faire doctrine in the 1850s and 1860s. Critical to American industrialization, this period witnessed the rise of coal as both an important fuel in manufacturing and as a mineral that required large-scale efforts at managerial and technological coordina-tion. American manufacturers also utilized wood, water, and animal power throughout the nineteenth century, but, as industrialization progressed, the use of mineral fuel expanded as well. In many ways the mining of coal served as one major predictor of industrial growth during the nineteenth century. This is why Benjamin Latrobe, as an acolyte of industrial development, en-thusiastically recorded his impressions of Virginia's early coal trade and why the subsequent growth of the coal industry of Pennsylvania and Virginia re-veals much about the early timing, nature, and location of the nation's In-dustrial Revolution.

Of course, individual actors such as Latrobe accounted for only a portion of the development of coal mining in the nineteenth century. State govern-ments played a critical role in developing the coal trade by constructing insti-tutional frameworks comprising a hodgepodge of policies. When considered in isolation, state-level programs such as internal improvements, geological surveys, and corporate chartering policies seem rather provincial, considering the grand march of nineteenth-century industrialization. In their aggregate form, however, these policies forged institutional pathways through which economic development occurred. Without these state-level frameworks for growth, natural resources could have laid dormant, entrepreneurs might have been stifled, and technological innovation would have found little succor. Politics, rather than nature, shaped the evolution of America's mineral fuel economy.[5]

To use a metaphor drawn from another form of nineteenth-century industrial power, consider the many millraces that dotted the countryside of early America. A well-constructed race eliminated the twists and turns of a natural creek and removed any obstacles in the waterway which might alter its flow. Waterways shaped in this fashion developed into a powerful stream capable of turning waterwheels and providing ample power. A poorly constructed millrace fails to create a powerful channel of water, as even a mighty stream can lose strength if it is forced to travel over rocks and thus split off in many different directions. In the nineteenth-century coal trade, entrepreneurship, skilled labor, and eager investment capital could lose momentum quickly if they encountered a poorly constructed framework for expansion. Similarly, a well-conceived institutional context for development amplified these forces of economic growth.[6]

The metaphor of a stream or a millrace is, of course, a simplified version of the interaction of many social and political structures during the nineteenth century. The chapters that follow present a more complex model of this process by comparing the ways in which legislative politics in Richmond, Virginia, and Harrisburg, Pennsylvania, constructed distinct institutional frameworks—the millraces of political economy, in a way—for the coal trades of their states. In Pennsylvania a distributive and flexible approach to economic policy making, though marked by corruption, partisan bickering, and regional animosities, created one of the most dynamic industrial regions in the United States. Policy making in the Keystone State assumed frighteningly inefficient dimensions, and initiatives for economic growth rarely emerged from the legislature in any kind of cogent form. In the end, however, distributive politics in Pennsylvania created an atmosphere conducive to dynamic industrial development. In Virginia participants in the coal trade encountered a system in which active, interest-driven politics operated within an institutional framework constructed to limit the flexibility of state government in order to preserve a stolid planter elite. Political institutions in antebellum Virginia, despite noble origins in republican virtue, simply could not facilitate disparate interests in the same fashion as their counterparts in Pennsylvania. Conservative interests of the East and the growing population of the West remained at odds with each other throughout the antebellum era, a struggle that stymied the growth of the coal industry in the state and demonstrated the reluctance of Virginia's government to accommodate economic change. Although both Virginia and Pennsylvania served as leaders in the early American coal industry, political forces in each state forged increasingly divergent paths as the nineteenth century progressed.

Each state created different economic environments, but the distinction should not be drawn between a "strong" Pennsylvania state and a "weak" state in Virginia. During the first half of the nineteenth century governors and state legislators across the United States aggressively pursued strategies for growth ranging from direct subsidies for valuable industries to the bestowal of distinct privileges to groups of entrepreneurs. Political controversies developed as a result, and many antebellum states followed a kind of "state in, state out" pattern in regard to more ambitious projects such as transportation networks. The infamous "canal boom" of the 1830s in states such as Ohio and Indiana provide a classic case for this model. When legislators in those states began to logroll and trade votes, nary an individual canal or road project suffered. When the smoke cleared, however, very few profitable projects existed and state debt ballooned. In the end, many historians argue, conventional wisdom regarding the role of the state shifted from an activist perspective to one embracing a more laissez-faire approach. In regard to the public underwriting of internal improvement projects, these arguments make sense. But does a state-in, state-out process in direct funding of economic projects reflect a diminished role for the state by the advent of the Civil War? Was the era of state activism confined to only a few antebellum decades?[7]

A broader examination of the institutional context of development created by state government suggests otherwise. Recent studies of government policy in the nineteenth century have replaced the idea of a bare-boned state of "courts and parties" with a more complex vision of actors and institutions sometimes working in tandem and sometimes working against one another. "Political economy" in this sense represents a series of political choices, often ad hoc bartering so beloved to nineteenth-century legislators which solidified into "policy regimes" that served as more or less permanent government institutions. Whether individual political decisions morphed into political institutions depended upon a number of factors, such as popularity, profitability, and convenience, which makes the formation of nineteenth-century political economy difficult to fit into formal models or formulas. A greater appreciation for the machinations of government institutions, even on a piecemeal basis, serves as a welcome change from the simple instrumentalism or pluralistic models of past generations.[8]

The role of slavery in industrial development offers another challenge for understanding the political economy of nineteenth-century states in the South. Most historians agree that slavery played an enormous role in shaping developmental paths of southern states, but they often fail to differentiate between the use of *slaves* in industrial endeavors and the impact of *slavery* upon a state's institutional framework. Instead, scholars have emphasized the powerful effects of slavery upon southern culture or the ability of southern manufacturers to use slave labor in their shops, forges, or factories. Perhaps this revisionism pushes the issue too far. Because slavery was so malleable, it appeared that the presence or absence of the "peculiar institution" might not be the culprit for the South's laggard industrialization. A wider view of the institutional context of industrial development offers a way to reconcile these apparent contradictions. The political framework of southern states demonstrates how slaves could engage in industrial pursuits quite profitably for their owners but that the wider impact of slavery upon a state's capital markets, political development, and economic outlook could result in stunted economic growth.[9]

Change within political institutions also affected the development of the millraces of state-level political economy. The period following the War of 1812 witnessed several dramatic changes, such as an expanded electorate, more equitable apportionments of state legislatures, and the emergence of a dynamic two-party system. The reification of racial hierarchies that severely limited African-American political participation, the emergence of professional lobbyists and party activists, and resistance to universal white male suffrage all occurred in important states such as Virginia and Pennsylvania and tempered the progressive nature of "Jacksonian Democracy." State legislatures, buffeted by both the challenges to the established political order and the counterrevolutions of intransigent elites, served as a battleground for competing visions of American representative democracy. They were not always up to this challenge. As one expert on political apportionment noted, state legislatures "were rarely portrayed or conceived as inordinately cerebral, deliberative forums in which the brightest or the most virtuous legislators convened to divine public policies best aligned with the common good." Or, as the editors of the Pottsville, Pennsylvania, *Miners' Journal* put it in 1825: "The practice of coughing and shuffling down empty talkers may perhaps be objectionable, but we think it far preferable to this tiresome trifling mode of popularity legislation."[10]

So it must be conceded that these millraces of political economy were not always constructed on sound principles and hardly offer a standardized model for historians to examine. State legislatures did, however, hold the preponderance of authority during a critical period of political and economic development in the United States. Individual entrepreneurs in the antebellum period constantly appealed to legislatures for direct bounties, special trading privileges, and many other forms of succor. The responses that they received varied. When Josiah White and Erskine Hazard asked for a corporate charter to improve the Lehigh River and develop the anthracite coal trade, the legislature was more than happy to provide them with the "privilege to ruin themselves." But, when iron makers asked for direct bounties for the use of Pennsylvania anthracite in the smelting of pig iron, their requests became bogged down in committee and ultimately failed. Much has been made of the "Market Revolution" and the critical role that business interests played in reshaping the contours of antebellum America. Less attention is paid to the state-level solons that in some instances aided entrepreneurial initiatives and other times stood squarely in the way of them. The fact remains that institutional pathways, no matter how shaky in their design or inefficient in execution, deserve a close analysis so that they may, in the words of one historian, "share the insight that governmental institutions can be agents of change," so that they may "help frame a more realistic, coherent, and inclusive account of the American past."[11]

How did these institutional factors influence the American coal industry in its formative stage? In Virginia the conservative polity empowered a particular vision of representation and governance. Eastern Virginians clung to their traditional power structure based upon two loci of power: a legislature proportioned upon the state's free and slave population and powerful local institutions staffed by traditional great landholders. These institutions, in turn, privileged the interests of the wealthy and powerful elite at the expense of small landholders, nascent manufacturing interests, and western Virginians. The message of economic policies forged in this system remained startlingly consistent: protect the value of landed property and promote the interests of agriculture. More often than not, but not always, the preservation of slavery rose to the forefront of this strategy. Coal mining operations in the Old Dominion, even though they utilized slave labor throughout the antebellum period, found themselves on the margins, if not completely ignored by their political leaders. As outsiders, they made various attempts to reshape Virginia's polity into a more equitable form but consistently ran up against an entrenched conservative leadership that fought tooth and nail against any reforms that might threaten the interests of an established agrarian elite.[12]

Pennsylvania's legislature similarly favored the well-propertied elites of its state. It was difficult, however, for a single set of interests to dominate a legislature proportioned by population and periodically readjusted to reflect changes in the geographical distribution of the state's population. When large landholders of the western mountains, iron manufacturers of the central counties, and financial interests in Philadelphia all complained that the legislature did not adequately serve their interests and privileged those of their competitors, they were all in a sense correct. The Pennsylvania legislature produced a cacophony of rival interests and competing factions throughout the nineteenth century. The most prevalent tune to emerge from the disarray in Harrisburg in the antebellum years, however, was an unbridled enthusiasm for economic growth. Opinions regarding how that growth was best promoted differed, of course, but policy makers hammered out these differences in both the formal debates of the House and Senate and in the informal realm of vote trading, logrolling, and outright corruption. The millrace that shady practices constructed in Harrisburg became an effective engine for the development of the coal industry. The policies that emerged from this system were by no means consistent, and yet, to the coal trade of both eastern and western Pennsylvania, they aided in the expansion of mining immeasurably.[13]

Using the coal industry as a barometer of regional economic change, I examine the relationship between public policy and divergent economic development in a northern and southern context to address long-standing questions of regional economic divergence in nineteenth-century America. The difference between these two states does not lie solely in the presence or absence of slavery or the structure of each government but in the interaction between these two critical components of political economy. Institutional structures provide the basic outline of the millrace, but the power of this construction draws from political issues such as the future of slavery in Virginia, the preservation of individual opportunity in Pennsylvania, and the rapid development of transportation networks in each state. Institutional frameworks and policy debates serve as inseparable elements in state-level political economy. But it is not enough to appreciate the presence of these state-level political economies; the careful scholar must also understand how they made an impact upon American industrialization. The chapters that follow seek to reassess one of the most enduring questions in the history of the United States, the nature of regional divergence, by telling the story of the rise of one of the most critical elements of U.S. industrialization, coal.

The Intersection of Politics and Geology

America's First Coal Trade

*I*n 1855 the former U.S. commissioner of patents and prodigious tech-
nophile Thomas Ewbank attempted to capture the essence of industrial dynamism in a book entitled *The World a Workshop; or The Physical
Relationship of Man to the Earth.* He described the reserves of American coal
as nothing less than heaven-sent. "A first element of progress for all time," he
wrote, "it is preposterous to suppose the supplies of coal can ever be exhausted
or even become scarce. The idea is almost blasphemous." As a reflection of
God's infinite benevolence, Ewbank argued, "the presence of coal—this is, its
power—is in almost every thing man makes, from a needle or penknife to a
steamship; from a brick to a city; a ball of thread to anything made of thread."
A decade later Eli Bowen reflected upon the providential nature of coal deposits in his work entitled *Coal and Coal Oil; or The Geology of the Earth.* Not
only did God place coal deposits under the plow, Bowen claimed, but also this

wealth was rendered "conveniently accessible" as a measure of "benevolent design" by the Creator, which demonstrated the "wise forethought and provision of a good father for his dependent and erring children." Although they came from different circumstances, both Ewbank, a former government official, and Bowen, a self-taught geologist of Pennsylvania's anthracite region, concurred that the massive reserves of coal situated within the boundaries of the United States was no accident but, rather, the work of a higher power.[1]

Whether the presence or absence of coal seams derives from providence, fortune, or geological arbitrariness is a question beyond the scope of this book. It is safe to say, however, that the successful adaptation of coal to manufacturing pursuits, perhaps owing more to human ingenuity than Ewbank and Bowen's idea of divine intervention, triggered the rise of a mineral fuel economy in the nineteenth century. The coal industry served many purposes in this regard. It straddled the breech between agriculture and manufacturing pursuits in many nations, served as an extractive enterprise that spurred technological innovation, and provided high-quality and low-cost fuel that unleashed the power of the steam engine. Historically, therefore, it seems only logical to assume that there is a high correlation between industrial development and the presence of coal mining.[2] Great Britain's economy, in particular, parlayed its generous coal reserves to create a dynamic industrial sector and emerged as the seedbed of fuel technology by the early nineteenth century. "Without coal," a British writer surmised in 1856, "no longer would our favoured country be the great factory for supplying the necessities of the great family of mankind." British canals, railways, manufactures, and exports all owed their dramatic growth in large part to the presence of coal, and the British system of coal-based industrial development provided a model for other aspiring nations. If some nineteenth-century descriptions of the universality of coal may have tended toward hyperbole, any assessment of the import of mineral fuel to nineteenth-century industrial life cannot be overstated.[3]

The U.S. economy, too, owed its industrial prowess to the ready availability of valuable natural resources such as coal. In the early years of its existence, however, the coal trade of the United States seemed quite modest. British colliers provided the initial model of a successful mineral fuel economy, but, as the British system moved across the Atlantic Ocean, the unique circumstances of the American setting created an altogether different coal industry. The idea of mining, transporting, and burning mineral fuel appears simple in the abstract, but many elements of the process vary according to the political, economic, and social circumstances in which they develop. In the bituminous coal region outside of Richmond, Virginia, the presence of slavery made it difficult for local colliers to transplant British methods of accessing, raising, and shipping coal, but their close proximity to tidewater allowed them to dominate the nation's coal trade for the first half-century after independence. Although Richmond colliers enjoyed the "first mover" advantage, certain limitations of their trade also became apparent as the demand for coal increased in the years following the War of 1812. Geology was very kind to colliers of Virginia during the early republic; the vagaries of plantation slavery were not as forthcoming. The history of the first coal trade of the United States demonstrates that, from the very beginning, coal could not be separated from its political context.

THE POWER OF COAL

What made coal seem so awe-inspiring, even spiritual, to nineteenth-century observers such as Thomas Ewbank and Eli Bowen? Since coal is essentially a rock that burns, one must begin with simple geology. In essence this mineral represents the captured energy of million-year-old sunshine. After plants die and decay, they release the sun's energy held in carbon, hydrogen, and oxygen compounds. When the encroachment of oceans buried the plant material of coastal swamps and marshland under layers of mud and sand millions of years ago, the energy collected by plants remained trapped in the decaying vegetation. This layer of rotting plant material, commonly known as peat, constitutes the building block of coal. As layers of sediment accumulate above, their enormous weight crushes the pores and cells of the plant matter, which drives out moisture and gaseous compounds containing oxygen and hydrogen. Heat, which is produced by forces beneath the earth's crust and the friction of rocks grinding together as mountains form, removes even more of the hydrogen and oxygen from the now ancient plant material. After millions of years of exposure to these geological forces, the mineral known generally as coal is formed.

Table 1.1. Coal Classifications Ranked by Carbon Content

Rank of Coal	% of Fixed Carbon	% of Volatile Matter
Meta-anthracite	98–100	0–2
Anthracite	92–98	2–8
Semi-anthracite	86–92	8–14
Low-volatile bituminous	78–86	14–22
Medium-volatile bituminous	69–78	22–31
High-volatile bituminous	69–50	31–50

Source: Adapted from Schmidt, *Coal in America: An Encyclopedia of Reserves, Production, and Use* 36–38.

Modern geologists classify coal by stages, or ranks, which are dependent upon the percentage of carbon (see table 1.1). Generally, a higher amount of fixed carbon in coal translates into more energy per unit. The various combinations of pressure and heat across the planet produced a wide variety of coal. In the early years of mining coal was rarely placed within this taxonomy and, instead, was known by the generic name of "sea coal," "pit coal," or "stone coal." Over the years, however, geologists broke down the mineral into a few major categories. Lignites, sometimes called "brown coal," contain relatively low amounts of carbon with a great amount of impurities, known as "volatile matter," which burn off when the coal is ignited. Bituminous coal contains less volatile matter than lignite and is quite common in the coal fields of Europe and the United States. Anthracites, ranked purest in carbon content, are usually found in areas that witnessed a great deal of mountain building, with its attendant heat and pressure. The types of coal most prevalent in the United States are anthracite and bituminous ranks, but beyond that it is difficult to generalize about American deposits because different seams contain coal of different chemical compositions.

Coal seams can range from a few inches to ten or twelve feet in thickness and can exist in very small areas or in fields that cover hundreds of square miles. Some seams lie parallel with the surface of the ground, while the vertical and horizontal upthrusts of mountain building can tilt other seams to intersect the surface at a 90 degree angle. Other natural forces, such as the changing course of rivers, erosion, or the movement of the earth's crust, alternately expose and hide veins of coal. One traveler to western Pennsylvania in 1792

made note of a "number of creeks and other waters, running for a considerable distance, over a bed of coal," and reported that "the banks of these western rivers are frequently lined with coal."[4] Natural exposure thus led to the
first discoveries of coal deposits and the initial stages of excavation. Some
mines are able simply to tunnel into the side of an exposed vein and follow it
along its course; others dig sloping or vertical shafts that intersect subter-
ranean veins. Whatever the method used to access the seam, colliers have always been subject to the whims of geology, and many a coal miner's fortunes
have been won or lost on the dips, breaks, and outright disappearance of coal
seams.

Aside from geological factors the productivity of mines also depended
upon the political institutions of the region. Because coal mining was a capital-
intensive endeavor, mining firms were among the first to utilize the financial
advantages of incorporation. This required a state willing to bestow such privileges to entrepreneurs and a legal system to protect them. Government officials throughout history also used geological surveys, bounties, and other
incentives to promote coal mining as a means to achieve economic wealth and,
by the mid-nineteenth century, military prowess. France, for example, threw
the weight of the state behind mining in the eighteenth century by offering
royal grants of land and the authority to mine coal to favored individuals. The
French also developed a national school of mines to train mining engineers
and promote theoretical research in the industry. Not all states promoted or
controlled coal mining as closely as the French, but the political contexts of
coal mining regions figure prominently in development. To argue that coal
seams themselves "have politics" pushes the analysis of the coal trade into the
realm of the absurd. That the *mining* of coal cannot be separated from its political context, however, seems to be an accurate assessment.[5]

England was the first nation to exploit its coal reserves extensively for use
in households and in manufactories. Abundant seams of bituminous coal,
commonly called "sea coal" because it was first discovered washing up on the
shores of Northumberland and the Firth of Forth, dotted the landscape of the
United Kingdom. Small-scale mining appeared in the coalfields of Scotland,
England, and Wales as early as the thirteenth century, and by 1306 London's

smoky atmosphere prompted a royal proclamation prohibiting artificers from using the mineral in their furnaces. The use of coal in growing urban centers nevertheless increased in Great Britain. During the 1560s miners raised more than a quarter-million tons of coal, of which London used about ten thousand tons in her hearths and furnaces. At first the use of mineral fuel was limited largely to areas containing easily worked outcroppings of coal. A timber shortage during the seventeenth century, spurred by naval construction and the growth of the charcoal iron industry, increased the demand for mineral fuel, which resulted in the expansion of the British mining industry. As wood grew costly and scarce and as mining became more extensive, British blacksmiths, bakers, and manufacturers of salt and glass began burning coal almost exclusively in their furnaces. By 1700 Great Britain's collieries raised 2.64 million tons of coal and had established the foundations for a "mineral fuel economy."[6] Although bituminous coal produced copious amounts of smoke, urban residents gradually made their peace with the sooty air. "Its supposed unwholesomeness was an objection," the American observer Benjamin Franklin wrote of Britain's experience with coal in 1785. "Luckily the inhabitants of London have got over that objection, and now think it rather contributes to render their air salubrious, as they have not general pestilential disorder since the general use of coals, when, before it, such were frequent."[7]

In addition to the increase in the consumption of coal, British colliers found innovations in extraction to be necessary after a few centuries of mining. British colliers first dug shallow trenches, or opencast mines, and also excavated large holes, known as "bell-pits" for their shape, in order to exploit seams close to the surface. Occasionally, miners opened a tunnel into a hillside to reach a seam, known as a "drift," or "day-hole." After a bell-pit endured prolonged exposure to surface weather or if a drift mine intersected an underground stream, however, coal workings filled with water, which needed to be removed in order to fully exploit the seam. Colliers tried to use gutters, drainage shafts, and various human- and horse-powered pumps to extract the water from their mines, but drainage remained a major impediment to expanding operations in the seventeenth century.[8]

The application of steam power to mine drainage in 1712 provided a novel solution to the problem and wedded this new source of industrial energy to the coal industry. The Newcomen pump, named after its inventor, Dartmouth ironmonger Thomas Newcomen, represented a breakthrough in mine technology. It essentially consisted of a coal-fired boiler, a condensing beam with a piston, a rocker beam, and a rod pump that used the partial vacuum of condensing steam to lift stagnant water from the mine to the surface. Although its early versions appeared awkward and inefficient, the Newcomen pump revolutionized coal mining by allowing miners to dig pits hundreds of feet deep and still remove water from the workings at a lower cost. In 1744, for example, Griff colliery in Warwickshire estimated that a Newcomen pump cost about £150 in fuel, maintenance, and labor, which replaced the £900 cost in feed and labor needed for a team of fifty horses. As improvements to the original Newcomen design, such as the one devised by the firm of Boulton and Watt in 1769, came into use, the cost savings from steam-powered pumps allowed mines to delve into seams of coal previously thought unreachable.[9]

A technological innovation in the production of iron further linked the coal industry to Britain's industrial prowess. As the nation's iron industry grew in importance, so did its dependence upon mineral fuel. Iron makers traditionally relied upon charcoal—made by charring wood—to fuel their furnaces. By the early eighteenth century, however, the shortage of wood had pushed up the cost of producing charcoal and in turn inspired a number of experiments aimed at utilizing mineral fuel in iron smelting. In 1709 Abraham Darby successfully adapted coal in iron making by using "coked" coal—coal with the impurities baked out—to smelt iron ore in his Shropshire furnace. Improvements in Darby's original process, along with the high cost of charcoal, eventually resulted in a massive shift among iron makers from wood-based fuels to mineral coke during the 1750s. The introduction of new methods of converting pig iron to bar iron paved the way for mineral fuel to replace charcoal in that secondary stage of iron production as well. By the 1780s mineral fuel dominated the British iron industry and immeasurably aided the nation's superior position in the Atlantic iron trade.[10]

In terms of structure British law determined that the right to mine coal lay with the owner of the surface land. Thus, with the geographical concentration of the industry, there were a number of large landowners who drew huge incomes from the mining of coal on their lands. The way that this coal was mined varied; some landowners developed collieries on their own, while others leased the land out to mining concerns for periods of two to three decades. The leasing system encouraged both sides to raise as much coal as possible, as colliers profited from maximizing output within the time constraint of their lease, and landowners earned more royalties as production increased. Yet certain large mining concerns sought to keep prices high. In the early eighteenth century the Grand Allies, a partnership of colliers dedicated to manipulating output and prices, emerged in the northeastern coalfields. Colliers in this region also founded the Limitation of the Vend, a cartel designed to limit shipments of coal to London and keep prices high. Economic historians estimate that this strategy raised coal prices by about 10 percent, even though members of the Limitation could not completely control the entry of new colliers in the Newcastle region. The cartel eventually broke down in the early 1800s, but the artificially high prices it had created allowed for substantial investments in new mining technology. The late eighteenth century saw some of the most dramatic reorganizations of the Newcastle coal regions, as increased investments in new pumping and raising methods occurred during this time of artificially high profits. By digging deeper shafts and exploiting more extensive holdings, Newcastle colliers entered the nineteenth century poised to capture economies of scale and thus earn large profits, despite the dissipation of the cartel.[11]

British mines of the eighteenth century employed a variety of underground and aboveground workers in order to raise coal. The most skilled of these were the "hewers," who supervised the extraction of coal at the face of the seam. In addition to hewers, British mines required other skilled workers to operate the carriage system, ensure safety underground, and operate machinery as well as a host of unskilled workers to cut, load, and haul the coal. Many of these jobs were unpleasant and dangerous, so mine operators often resorted to bonds to keep their labor force in place. The bond was an annual contract that defined the rate of payment, obligated the miner's family to keep company housing in good shape, and set up an arbitration process for the resolution of disputes. As local law officials enforced bonds with the threat of imprisonment, these agreements served to keep any restless workers in place for at least a year and often included no-strike clauses. Despite the restrictions, mine workers generally favored the bond system, especially when they could force mine operators to cough up "binding money" to induce them to sign during a tight labor market. By the early nineteenth century binding evolved into a standard practice in the labor relations of the British coal industry much like the lease agreements between landowners and mine operators, which contributed to a relatively stable industrial community that could implement technology without fear of bankruptcy or a disruption in supply or demand.[12]

The fuel needs of British manufacturing by the 1700s thus initiated the first large-scale attempt to harness the power of coal. The cost efficiencies of coal-fired steam engines and coal-based iron making improved over time, moreover, and coal emerged as a necessity, not simply a substitute fuel, in British manufacturing. With the tacit approval of government officials, colliers in the Newcastle region organized to regulate production in order to implement new technology and ensure price stability. As a result of technological and political innovations in the coal trade, British iron, along with its textiles, eventually dominated markets worldwide. Coal also warmed the hearths of British cities, as anyone inhaling London's distinct atmosphere could testify. Simply put, mineral fuel emerged as an indispensable part of British economic, social, and political life over the course of the eighteenth century. As the British colonization of North America progressed, one would expect that the rich deposits of coal in the "New World" offered a way to expand this coal trade across the Atlantic.

Barriers to the development of coal mining in the North American colonies, however, were significant at first. Familiarity with Newcastle coal led many colonial consumers to import it across the Atlantic rather than exploit local coal reserves. As late as 1770, Britain exported over 8,700 tons of coal to its colonies, most of it for use in growing coastal cities. Boston, for example, received 701 tons of coal in 1771, three-quarters of which came from British sources. The major colonial destination for British exports, New York and Philadelphia, received 2,474 and 1,241 tons, respectively, that year. In both markets British coal constituted about 90 percent of the shipments. When imported coal could not be had, urban colonists drew upon local stocks of wood hauled in from surrounding areas. Although some Americans mined coal for local use and a small domestic trade was present in the Richmond basin, forests and imported coal served as the main source of domestic and manufacturing fuel throughout the colonial period.[13]

Following independence, a growing commercial rivalry between Great Britain and its former colonies did not eliminate shipments of coal to the United States. Wood and waterpower still dominated the energy markets of the early republic, but the utility of mineral fuel gained ground in urban markets. Blacksmiths, glassblowers, and bakers often used British coal when wood became expensive, and, indeed, most early Americans seemed content to import their mineral fuel in the first few decades of independence, even though American patriots offered critiques of the continued use of British coal. In 1790 Thomas Mann Randolph of Richmond wrote to Thomas Jefferson that, although the bituminous coal of Virginia "contains a greater quantity of inflammable substance than any I have ever met with," it was unfortunate that importation of British coal "had not been so much discouraged as to induce moneyed men to lay out their capitals in that way." That same year nearly eight thousand tons of Newcastle coal found its way into American ports.[14]

In the cities along the eastern seaboard such as Philadelphia, domestic consumers began to take notice of coal as a source of heat, even though Americans of the late eighteenth and early nineteenth century were more likely to burn wood than coal to warm their homes. The *Literary Magazine* published a letter in 1804 which declared regular coal shipments from Virginia to be a necessity for future urban growth as local stocks of wood continued to dwindle in seaboard states. "From the situation of these states it appears necessary that some mode of procuring coal," the author claimed, "or some other substances for the consumption of fire, instead of wood, ought immediately to be pursued." Even rural consumers found the continued reliance upon wood fuel to be problematic. "The increasing scarcity and dearness of firewood indicates the absolute necessity of attending in the future to the coal mines of this country," wrote one Philadelphian in 1790, "especially with a view to our capital towns." Samuel Breck, writing from his estate on the Schuylkill River in the winter of 1807, observed that "wood is enormously dear, and my farm does not afford me any." He traveled to the city in order to purchase an iron grate used for burning coal, noting, "I mean to use that fuel if it answers my expectations."[15]

Consumers of manufacturing power also began to recognize the power of coal during this period. In 1784 a Philadelphia blacksmith advertised for a "fireman" who "understands the London way" of burning coal. A correspondent to the *American Daily Advertiser* argued that the consumption of wood by brewers, brick manufacturers, potters, and soap makers raised the price of fuel for everyone in the winter of 1804. "Coal would answer the purpose for these artists as well as wood," he claimed, "and no doubt would be used by them, provided coal could be had as cheap or cheaper than wood." It is difficult to assess exactly how many small-scale manufacturers shifted from wood to mineral fuel, but qualitative evidence from periods in which coal was difficult to procure suggests that many artisans were dependent upon coal by the War of 1812. In the summer of 1813, for example, two Philadelphia merchants implored Harry Heth of Chesterfield County, Virginia, to ship bituminous coal to them. They maintained that "we have not a doubt but a Cargo of good coal would readily sell for 90 to 100 cents per bushel," an estimate that tripled prewar price levels. "We are selling them out of the yard at the latter price," they noted, before offering implicit instructions on how to keep coal vessels speedy and elusive, "should you conclude to risque a load shall do all in our power for your interests & benefits." The high demand for coal during the summers of 1813 and 1814, a period when domestic consumption was at its lowest, triggered a full-blown fuel crisis among blacksmiths, bakers, and other small urban manufacturers.[16]

The consumption of mineral fuel at home clearly had a potential for dynamic growth during the early republic. Although urban residents initially were quite willing to burn foreign as well as domestic coal, the underlying assumption among most Americans was that their young nation's infant coal industry should eventually supplant the transatlantic trade. Economically, the creation of a vibrant industry patterned on the British model made sense. Consumers of fuel would benefit from lowered prices, and American shipping and carrying trades would also grow along with the domestic coal industry. The prospective value of mining coal to the U.S. economy was not the only argument in favor of a domestic trade. A series of political developments during the early republic shifted the young nation's mineral fuel economy away from the transatlantic import trade and toward a more local outlook.

Early policy making by federal authorities, for example, suggested that imported coal would not be welcome in U.S. markets. Although Federalist policies sought to cultivate strong economic ties with Great Britain, opposing political leaders fought to sever, or at least minimize, this relationship. The rise of Thomas Jefferson and the Republicans to power in 1800 signaled an ideological shift in federal policy making which focused less upon cultivating imports and looked, instead, toward expanding the economic horizons of the United States westward. Although important developments during the period of Republican rule such as the Louisiana Purchase and the various embargo acts had little to do with the coal trade, the importation of British coal suffered tremendously in both practical and symbolic terms.[17]

Individual states pursued a more active approach to economic development than the federal government during the early decades of the nineteenth century and often focused their energies on internal state matters, which limited the prospects of British coal in the U.S. economy even farther. As the Jefferson and Madison administrations struggled to reconcile the need for a national system of roads, canals, and water improvements with their perceptions of republican virtue and the more pressing constraints of Republican frugality, state governments in New York, Pennsylvania, and Virginia led the effort to develop local resources by subscribing aggressively to local internal improvement initiatives. These "developmental corporations" offered an ideal way for public officials to promote economic growth and, in the words of historian John Majewski, "enrich stockholders less through dividends than by bringing the fruits of market development to local communities." Local coal reserves, along with rich deposits of iron, copper, and other minerals, materialized into assets of great significance to state-level policy makers during this period.[18]

Federal protection sounded the real death knell for foreign coal and provided domestic producers with breathing room. British merchants faced prohibitive American tariff barriers and found the export trade in coal to the United States quite limited by the early nineteenth century. In 1808 Albert Gallatin reported to the U.S. Senate that Virginia coal sold at about twenty cents per bushel in Philadelphia, which earned carriers a decent profit. British coal, on the other hand, cost eighteen cents per bushel at Liverpool and paid a duty of five cents. When merchants figured in the cost of freight and insurance, British coal had to sell for at least thirty cents a bushel to net a profit.[19] Following the punishing embargoes of 1807–8, hostility toward British imports, the desire to encourage domestic production, and the need for federal revenues boosted tariff levels to ten cents a bushel in 1812—about 15 percent of the wholesale price of British coal.[20] This relatively high tariff, combined with the decline in imports during the War of 1812, provided a window of opportunity for the U.S. coal trade which was, in the eyes of one historian, "equivalent to extreme protection." The tariff on coal dropped to five cents a bushel again in 1816—which ranged between ten and 25 percent of the price of foreign coal in New York—and remained at about that level until 1842. British imports managed to bounce back after 1815, but they never again exceeded more than 10 percent of U.S. production. Tariff levels thus did not completely push British coal out of American markets, but they did severely restrict its ability to compete with the domestic product. After the War of 1812 British coal never again played an important role in U.S. markets, even though the image of cheap Newcastle coal pouring into the United States remained one of the most enduring leitmotifs for American protectionists of later decades.[21]

Constraints on trade and high tariff levels not only undermined British imports during the War of 1812, but they also spurred the growth of a domestic American coal trade patterned on this British model. As a low-value, high-bulk commodity, coal was unlikely to serve as a major trade good in the transatlantic economy in the long run, and political circumstances exacerbated this tendency. American policy makers could draw upon the British example for inspiration and adaptation, but they began to look inward to encourage the exploitation of domestic coal seams in the United States. Despite initially low levels of production, economic nationalists abhorred the importation of a commodity that seemed to be readily available at home. "Our resources are so abundant, that we shall be happier in procuring fuel from our own mines, than deriving it from yours, subject, as it ever must be, to the impositions, interruptions and vexations of a foreign commerce," one patriot wrote in 1811. "Moreover, our domestic coal will be brought to the consumer cheaper than yours. And you know, where fuel is cheap, all the arts are carried on with a corresponding thriftiness and ease." Domestic coal did not yet serve as the primary fuel of American hearths, forges, and furnaces, but the presence of a coal trade became an important component of economic nationalism in the early nineteenth century. If they would not import the coal itself from Great Britain, policy makers surmised, the replication of Britain's mineral fuel economy could prove quite lucrative.[22]

THE OLD DOMINION AND THE NEW INDUSTRY

When political economists discussed the importance of developing of a domestic coal trade, all eyes turned toward Virginia. Just as the Old Dominion's tobacco enriched colonial society, many American policy makers expected its coalfields to provide a measure of the self-sufficiency in raw materials which was so critical to the future of the young nation. In 1787 Thomas Jefferson described the Richmond area as "replete with mineral coal of a very excellent quality" and noted that a number of proprietors opened coal pits that "before the interruption of our commerce were worked to an extent equal to the demand." In his famous *Official Report on Publick Credit* Alexander Hamilton called Richmond coal an important national asset as a domestic and industrial fuel. An early American political economist, Tench Coxe, expected the collieries along the James River to provide inexpensive coal so that British coal would "be rendered a very losing commodity" and hoped that the coastwise trade would act as a "valuable nursery" for American seamen. Albert Gallatin used the increasing significance of the Richmond basin's coal to justify his ambitious plan of a national system of roads and canals. Of Richmond's coal he claimed, "the quality is excellent for all manufacturing purposes, and if properly selected, equal to any foreign coal for domestic uses." Escalating fuel prices in American cities would "soon create a dependence on it for both purposes," Gallatin noted, "and it is daily so much increasing in use, that it must command an immense carriage in which a minute saving of expense will be of the utmost consequence." Whether Virginia's coalfields would provide sufficient fuel for all of the young nation's energy needs was not the question—wood and waterpower still dominated fuel markets. But, as the economic nationalism of the early republic came into focus, Virginia's coal reserves contributed significantly to ideas of self-sufficiency and future growth.[23]

Although policy makers usually meant the Richmond area coalfield when they referred to "Virginia coal," mineral reserves in the Old Dominion were scattered across a variety of geographic regions. Mountains split antebellum Virginia into two distinct eastern and western sections, each composed of two major geographical subsections. The area of coastal plains from the Chesapeake to the fall line of Virginia's eastern rivers is known as the tidewater. Being the earliest area settled by Europeans, the tidewater was home to some of the largest and most established plantation families in the Old Dominion. By the advent of the nineteenth century the influence and profitability of this tobacco-growing aristocracy had waned, but tidewater planters nevertheless continued to exert a great deal of influence in Virginia politics and culture. The piedmont constitutes the rolling hills of fertile soil from the tidewater region to the Blue Ridge Mountains. In the early nineteenth century the piedmont region emerged as the breadbasket of Virginia. Planters there grew some tobacco, but they also profitably raised wheat, corn, oats, and potatoes to make Virginia's agricultural economy one of the richest and most diverse in the United States. The piedmont gives way to the valley, a hilly and agriculturally rich region hemmed in by the Blue Ridge to the east and the more imposing Allegheny Mountains on its west. The residents of the valley tended to live on smaller farms than their counterparts in the piedmont, but they nonetheless prospered by raising wheat, rye, barley, and hay. Finally, the transmontaine, or trans-Allegheny, region constitutes much of present-day West Virginia—an area characterized by rugged hills, numerous creeks and streams, and stony soil. At the time of the American Revolution, residents of this sparsely populated country raised grains on small patches of cleared land and participated in a thriving livestock trade.[24]

Like its diverse geographical regions, the Old Dominion's coal reserves were divided into eastern and western categories. Virginia's large eastern bituminous reserve, commonly called the Richmond basin, lay practically in the shadow of the Old Dominion's statehouse and had been worked since the early eighteenth century. The Richmond basin was a diamond-shaped coalfield of about 150 square miles, stretching a little over 30 miles from Goochland and Henrico counties in the north to Amelia county in the south, with its widest point being about 10 miles across, in Chesterfield County. The James River runs through the diamond in its northern section, and the area was dotted with a number of small creeks. These creeks cut into the land so that the veins of coal lay very close to the surface, sometimes breaking out to create outcroppings of coal apparent to the naked eye. "In several Places is Coal near the Surface of the Earth," one early description of Virginia claimed, "and undoubtedly in time they will either have Occasion to Vent for it, to supply other Places, if they will not use it themselves."[25]

French Huguenots discovered coal in the Richmond basin around 1699, and the early coal pits, as they were called in the British fashion, probably began operation in the 1740s. By midcentury merchants carried Richmond coal to colonial seaports, and in 1767 Samuel DuVal proudly declared his coal from the Deep Run pits on the north bank of the James as "equal to Newcastle coal." Coal mined in the Richmond area was bituminous with a carbon content that averaged about 75 percent, putting it in the "medium-volatile" rank. For American consumers already familiar with British coal, Richmond bituminous served as an acceptable substitute. A blacksmith who had previously burned imported coal could easily use Virginia bituminous in his forge, and Virginia coal assumed a small share of the coastwise market. In 1763, for example, Richmond basin colliers shipped 102 tons of coal to Philadelphia, 232 tons to Boston, and 247 tons to New York City.[26]

In addition to Richmond coal, Virginians had long been aware of the abundant coal in their western counties. Virtually the entire area of present-day West Virginia contains seams of bituminous coal. "In the western country," Thomas Jefferson noted of his native state in 1787, "coal is known to be in so many places, as to have induced an opinion, that the whole tract between the Laurel mountains, Mississippi, and Ohio, yields coal." Early visitors often marveled at the thickness of coal beds along the Ohio River, and residents of Wheeling began to burn coal from local pits in 1810. Deposits of bituminous coal in western Virginia defied any systematic mapping until the mid-nineteenth century, but economic activity seemed to spring up wherever seams broke the surface of the ground. By the late 1820s, for example, glass manufacturers had situated themselves next to a coal outcropping along the Ohio River, where they could mine local coal for as little as one cent a bushel.[27]

At the same time, an early manufacturing demand for coal developed in the Kanawha River Valley, where production of salt for western markets ben-

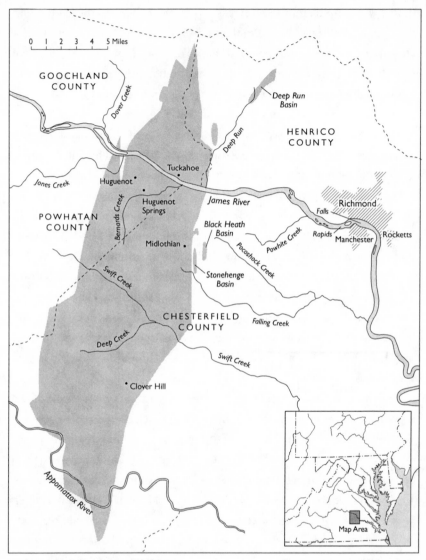

Map 2. The Richmond coal basin

efited from the presence of cheap bituminous fuel. Early manufacturers of salt used large kettles to boil saline water in order to remove impurities and reduce the salt to a crystal form. As the industry expanded, the constant firing of the salt kettles quickly reduced surrounding stocks of timber, and the price of wood fuel increased. In 1817 John P. Turner opened the Kanawha Valley's first coal mine for the purpose of providing fuel for salt manufacturers, and Colonel David Ruffner first used coal to fire his kettles two years later. A symbiotic relationship thus developed between coal and salt in the Kanawha Valley, as local salt producers quickly became familiar with the area's many workable seams of bituminous coal. Less was known about the exact extent and quality of coal in this region than that of the Richmond basin, but local residents and visitors alike nevertheless marveled at the abundance, easy accessibility, and apparent inexhaustible nature of mineral deposits in the mountains of western Virginia.[28]

The mountains also influenced the social and political context in which the Old Dominion's coal industry developed, as the division between eastern and western Virginians assumed ethnic, religious, and racial dimensions. White residents of the tidewater and piedmont regions tended to be of English descent and Anglican faith, and they largely observed the deferential political culture of colonial Virginia. The whites of the valley and transmontaine regions of western Virginia, on the other hand, were more likely to be of German or Scotch-Irish descent and observants of the Lutheran or Presbyterian faith, and they tended toward a distrust of hierarchical forms of authority. Westerners, often referred to as "Cohees," a derivative of the pietistic expression "quoth he," resented the influence and prestige of their eastern counterparts, or, as they called them, "Tuckahoes," a moniker adapted from a Native American term for a swamp root. But the most striking difference between east and west in the Old Dominion lay in each region's relationship to slavery. In the early nineteenth century the tidewater and piedmont regions supported a thriving agricultural economy based upon the labor of enslaved black Americans. The majority of white families east of the Blue Ridge owned slaves, and slavery was, therefore, much more entrenched in the tidewater and piedmont. In fact, 93 percent of the Old Dominion's 345,796 African-American slaves lived east of the Blue Ridge in 1800. That same year eastern Virginia accounted for 76.6 percent of the Old Dominion's total population of 860,046, but nearly half (48.9%) of those Virginians were enslaved. As a result, eastern Virginia's share of the free white population of the Old Dominion fell to 65.4 percent. In contrast, only 13.3 percent of the 201,558 western Virginians in 1800 lived in bondage.[29]

Political developments in Virginia during the early republic exacerbated these long-standing sectional tensions. Virginia's constitution of 1776, despite its birth in Revolutionary circumstances, produced a polity deeply rooted in traditional institutions which barely revised the county-based governing structure of colonial Virginia. Power largely resided in the hands of the General Assembly, composed of an upper and lower house. The House of Delegates consisted of 2 representatives from each of Virginia's sixty-two counties and 1 each from the towns of Williamsburg and Norfolk—126 members in all. Virginia's upper house, the Senate, consisted of 24 members, each elected from geographically fixed districts, regardless of their population. This geographical, as opposed to population-based, apportionment favored the older tidewater region, which had more counties because British colonists had settled the tidewater earlier than the other regions of Virginia.

Virginia's constitution also preserved the power of the county courts, which had served as the major governmental units of the colonial era and remained, in the words of one historian, "self-perpetuating family oligarchies." The governor appointed county justices of the peace from the local landowning elite. These justices of the peace, who administered over the county courts, held authority in all criminal and civil suits. County court officials also held roughly half of the seats in Virginia's House of Delegates during the early nineteenth century, which often blurred the line between county and state-level decision making. A provincial, county-oriented system thus survived the radical political changes experienced by many states during the American Revolution and set the Old Dominion's political and legal institutions upon a decidedly conservative course. The county court remained one of the most powerful institutions in Virginia politics throughout the nineteenth century.[30]

With their control of local politics firmly entrenched by the county courts, elite slaveholding families exerted their influence over statewide policies through geographically based representation. In this sense Virginia's legislature used a system similar to the "rotten boroughs" in England for the first three decades of the nineteenth century, as each county sent two representatives to the General Assembly regardless of population. With nearly twice as many delegates in the lower house, from which all legislation originated, slaveholding elites exercised their authority in Richmond without significant opposition. Thus, at an early stage the Old Dominion placed ownership of physical and human property at the center of politics. Voting privileges in Virginia amplified the political power of property even more, as suffrage remained limited to freeholders, still defined by an act of 1736 as a white male who held one hundred acres of unimproved land without a house or twenty-five acres of improved land with a house. The General Assembly reduced the freeholding requirement on unimproved land to fifty acres in 1785 and made no significant revision to that provision until 1851. The freehold requirement for voting favored the interests of large landowners in Virginia's electorate, as well-propertied citizens could vote in every county in which they had a freehold.[31]

Structural elements of Virginia's constitution atomized political power across the state but generally gave the tidewater-dominated legislature the preponderance of influence in state government. Virginia's governor, for example, was elected by the General Assembly and exercised no veto power over legislation. Other than control of the state militia, the power to issue executive pardons, and the ability to appoint judges, the governor had very little authority in Richmond. The executive's nominating power might have been important, given the power of county courts, but tradition and custom limited the choice of county officials to the local gentry, and a joint ballot of the legislature elected all state judges before they received their lifetime commission from the governor.[32] With the majority of delegates drawn from their districts and with such a weak executive branch, tidewater elites were thus guaranteed control over all three branches of the Old Dominion's government. Finally, the 1776 constitution provided no means for reapportionment of the legislature or for amendment, so any significant revision of Virginia's state structure required the consent of the majority of the same body that benefited the most from its inequalities.[33]

This distinct political arrangement provided the institutional framework for the nation's first large-scale coal trade. The structure of Virginia's earliest coal industry, not surprisingly, reflected the predominance of agricultural pursuits in the Old Dominion's social and political landscape. Local planters owned the land that encompassed most mining operations in the Richmond basin, but, unlike their British counterparts, many Virginia colliers entered leases of only a few years, with the intent of raising as much coal as possible in a short period of time. The owners of the Tuckahoe Pits, for example, lured potential lessees with the idea that upward of four hundred thousands bushels of coal could be raised on their land in a year. This short-term strategy and the lack of regular transportation along the James River resulted in the stockpiling of large amounts in the immediate area of the pits. In March 1815 Thomas Taylor publicly announced the sale of two to three hundred bushels of coal lying at the edge of a navigable creek some ten miles from Richmond. If they did not get buyers soon, Taylor lamented, the coal would have to be put up for public auction. Rather than providing regular shipments down the James River and from the tidewater port there to the cities of the eastern seaboards, most Richmond-area coal miners responded to orders placed by urban merchants. As a result, timing one's annual "harvest" of coal—Virginia colliers often referred to their product in such terms—and its arrival to market was critical.[34]

Novice colliers dominated the early history of the Richmond basin. In 1796 the French Duc de la Rochefoucauld visited the mines of two Richmond-area colliers and described them as "gentlemen who are neither chemists nor mechanicians, are content to grope their way without applying for advice to more enlightened men; for there is not one person throughout America versed in the art of working mines." In order to highlight the ease of extraction for amateur miners, advertisements for coal lands mentioned the mineral's proximity to the surface, along with the ubiquitous depiction of the region's supply as "inexhaustible." The short-term investment and lack of practical expertise among many coal pit managers led to a method of mining known as "trenching." Rather than tunneling alongside or into the vein of coal, the trenching method resembled more of an open quarry. These open-air diggings measured up to thirty feet deep and often filled with water before they could be fully worked. Although some early Virginia coal entrepreneurs experimented with deeper shafts and water pumps to keep the mines dry and maximize their mine's yield, most miners abandoned their works when it became too expensive or too soggy to remove coal. Abandoned pits and tunnels on one's land were a regular hazard for Virginia's early colliers. Samuel Paine wrote to his partner in 1801, "We are sinking a pitt at the extreme point of the commence-

ment of the old works and if we escape the water of the old works, I see nothing to prevent our doing as well as we have a right to expect for another year."
Haphazard and short-term planning placed serious constraints upon the expansion of the region's production.[35]

Because Virginia was a slave state, Richmond area colliers used slave labor
in their mines. At the time British miners used the bond in order to secure labor, but slavery presented a whole new set of challenges for miners in Virginia.
On the eve of the Civil War, for example, mining firms both owned and leased
black slaves year round, but in the early nineteenth century slaves were employed exclusively in mining only rarely. The plantation visited by Duc de la
Rochefoucauld in 1796 used five hundred slaves but alternated its work as-
signments between agricultural and mining pursuits. Miners often relied
upon a local surplus of labor to augment their workforce. The hiring or rent-
ing of black labor was not always successful, as owners preferred to engage
their slaves in agricultural work. Richmond basin producers responded to this
problem by purchasing slaves of their own to employ in their mines alongside
leased bondsmen, but most colliers still found it necessary to hire the majority of their slaves from other owners. These occasional shortages of labor thus
exacerbated the disorganized nature of the early Richmond basin coal industry.[36]

Given the conditions of coal mining in the Richmond basin, labor shortages appeared inevitable. After descending a few hundred feet into the earth
by way of a tub-like container called a "corve," workers trudged along a damp
and narrow shaft to get to their particular gallery, a small room that jutted into
the coal seam. Although thick walls of coal and rock, called "breasts," framed
each gallery, a constant creaking and groaning reminded miners that the roof
of their gallery could shift or even collapse at any moment. Later in the region's development, mine operators left only pillars of coal standing to support the roof and expected to remove, or "win," the coal from them by allowing the roof to collapse by design. "A strong sulphurous acid ran down the
walls of many of the galleries," according to one observer in 1818, and water
constantly seeped into the work areas. After negotiating a maze of tunnels in
near darkness, miners arrived at their particular gallery with no small sense of
relief. Awaiting them were hours of backbreaking work. With a pick and

shovel, miners tore at the coal seam, loaded carts with the broken coal, and then transported it to the surface. No wonder that in 1812 a slave named Jeffrey "came on the Monday morning saying he did not like to be at the coal pitts," to his master, who recommended one hundred lashes as an example for other reluctant miners.[37]

Workers faced even more problems in eastern Virginia's mines. The bituminous seams in the Richmond basin held great amounts of methane gas, called "fire damp" by miners, which added the threat of explosion or suffocation. Miners in Great Britain and Pennsylvania eventually developed sophisticated ventilation systems to remove hazardous gas, but Virginia miners used relatively crude and potentially fatal methods. Richmond miners sometimes employed a "firing line," consisting of a candle on a string attached to a pulley on the coal's face which was pulled toward the gassy area. Another approach was to have an unlucky worker wrap in a wet cape and inch his way toward the methane pocket; once in position, this "cannoneer" would raise a lit torch into the gas and ignite it. If a fire started in the mines, it could smolder for months or even years. In 1810 the manager of one mine descended into a shaft to appraise a fire but "had not been down five minutes before the hands began to stagger & fall from the effects of the Sulphur, which is intolerably strong." Lingering fires could ignite pockets of methane exposed by miners, as in 1839 when an explosion ripped through the Black Heath pits, killing more than forty workers and etching the blueprint of the mine's tunnels and galleries on the surface.[38]

Poor transportation from the mines to shipping centers also plagued the early development of the Richmond basin. Local roads had been in place for years, but they were little more than crude paths and could hardly handle the heavy traffic of a growing coal trade. Turnpike roads provided one solution to this problem. An appeal by citizens from Henrico County for a turnpike charter to the legislature in 1803 made clear the dilemma. When the Deep Run coal mines increased production, they explained, "the Deep Run and Westham Road were so cut with waggons as to be almost impassable at some seasons of the year for Loaded Waggons and Carriages." Existing roads could not accommodate wagons that carried heavy loads, which meant that a mine's shipment had to be stretched out over more wagons and more trips to keep roads navigable. A regularly maintained turnpike, it was believed, could handle an increase in coal traffic.[39]

In Chesterfield County the Manchester Turnpike Company (chartered in 1801), a thirty-foot-wide road that connected the ferry landing at Manchester with the Falling Creek area, attempted to redress the inadequacy of county roads. The turnpike counted at least three area coal mine owners among it seven directors.[40] The toll structure of the road did not differentiate between coal and other commodities, but coal wagons had to pay an additional fee of one-half of the toll on the return trip to Falling Creek. In 1804 the legislature amended the firm's charter and required the directors of the road to "take such a direction as would render it as convenient as possible to the different coal mines in the vicinity of Falling creek" and to "strike the Black Heath coal pits road" so as to increase the regular transportation of coal.[41]

Shipping coal to Richmond by water provided an attractive alternative to overland travel, for the Richmond basin is crisscrossed with small creeks and rivers. Unfortunately, the majority of these natural waterways were very shallow and thus unreliable for regular navigation. Spring freshets created by melting ice and temporary dams helped, but sometimes water levels remained so low that coal piled up at the mine's mouth for months at a time. The James River itself was notoriously unpredictable for cargo traffic. Orris Paine, who managed a coal mine in Louisa County, knew that the difference between profit and loss often depended upon the fickle water route to Richmond. "If no interruptions take place in gitting coal down," he wrote to a business partner in 1797, "we shall look like men, that are independent." In the fall of 1801 Paine complained that "the extreme lowness of the river prevents our getting the coal down"; "could we have got down what we have raised this year, we should have derided something handsome."[42]

The successful use of waterways to transport coal in eighteenth-century Britain provided one potential model for the future of Richmond's coal trade. British colliers estimated that the overland transportation of coal doubled its price in ten miles, whereas water routes could carry a coal twenty times farther than land transport at exactly the same unit cost. British coal usually traveled a short distance by wagon from the mine to the "staith," a kind of riverside quay that guided coal from the wagon in to the hold of a vessel moored below. This vessel, called a "keel" in the British trade, then transported the coal along a river or canal to market. In 1771 Jonathan Williams marveled at the duke of Bridgewater's integrated system of canal transportation which "brings out his Coals in a very easy Manner." Whenever possible, British colliers shipped their product via river, canal, or coastal vessels.[43]

Artificial navigation along the James River had been under way in Virginia since the formation of the James River Company in 1785, and by the early nineteenth century the firm had constructed a series of dams and channels to improve navigation along the James. The company also cleared major obstructions in the river and completed a short canal that linked the bituminous coal mining region with Richmond. This provided the best water-based transportation for colliers, and in 1803 over two thousand coal boats passed through the James River Company's navigation.[44]

But these efforts did not necessarily solve the Richmond basin's problems with reliable coal transportation. In fact, the James River Company's works disappointed the coal interests when many heavy barges could not make the journey during the dry season, as water levels in the JRC's navigation dwindled to less than one foot in depth. Some boats even dumped coal overboard in order to pass through the shallower areas of the James. In 1804 colliers of Powhatan County complained to the legislature that the delay in deepening the artificial navigation of the river was injurious to the area's coal trade. The James River Company "tends to sacrifice the interests of the community," they maintained, "and to promote that only of the individual share-holders & the inhabitants of the city of Richmond." Early transportation networks thus made the regular shipments of coal from the Richmond basin mines a complicated endeavor.[45]

Weather conditions, gluts on the market, and other unforeseen circumstances made it difficult for coal dealers to ensure a steady flow of sales even in the best of times, and Virginia colliers struggled to address the unique conditions of the fuel market during the embargo and war years. For example, in 1811 Harry Heth received requests from merchants in Baltimore and New York to send as much coal he could manage. Later that fall Washington coal broker John Davidson wrote, "You are now in the full tide of a glorious harvest, with more demand for your Coal, than you can satisfy." But he also warned, "It was not always so, and may not be the case in the future." Davidson then ended his missive with a portentous note: "The Richmond Coal does not answer well here; but if I can get no other without waiting too long I must take what I can get." Heth continued to receive orders for his coal during the War of 1812. The war eventually cut into Heth's ability to fill these orders and hurt his business. "This d——d war has all but ruined me," he confessed to a friend, but he also predicted that during peacetime his profits would be great. By the close of hostilities Heth's prediction had proved correct, and he received a barrage of requests for his coal.[46]

Harry Heth's struggle was typical of many of Virginia's early colliers, and his career serves as an excellent microcosm of the Richmond basin's relative decline. A slave owner who raised wheat and corn as well as mined coal, Heth owned a large plantation on the outskirts of Richmond and was as wedded to the institution of slavery as any tobacco planter in the region. In other respects this former Revolutionary officer and tobacco agent parted with his peers in the tidewater. As the son of an English immigrant, he cultivated contacts with British colliers at the same time that he lived and socialized among his fellow planters. Throughout his career as one of the Richmond basin's leading colliers, Heth attempted to adapt the hard-won wisdom of British mines to the particular contours of eastern Virginia's coalfields. Although his family remained in coal mining long after his death, Harry Heth discovered that the political and economic institutions of the Old Dominion placed severe limitations on his chosen enterprise.[47]

During the postwar years of great demand, problems arose that began seriously to impede Richmond coal's ability to compete in urban markets. As demand for their coal grew in Boston, New York, and Philadelphia, Heth and other producers simply could not consistently fill orders for high-quality coal. Even Harry's nephew William, who reported optimistically from New York in March 1815 of the need among blacksmiths, bakers, and domestic consumers for his uncle's product, was unable to procure enough coal from Heth to break even. As urban consumers sought to increase their consumption of coal from the Richmond basin, miners failed to increase the amount of quality bituminous coal to meet demand. When one considers that the Richmond coal basin lies only a few miles from tidewater and that merchants carried Richmond coal to American cities beginning in the early eighteenth century, this failure appears puzzling.[48]

The inability of Richmond basin miners to capitalize on the favorable market conditions both during and after the War of 1812 demonstrated the problems of engaging in an industrial pursuit while situated in a region without an economic infrastructure geared toward expansion. In order to meet postwar demand for their coal, for example, Richmond basin colliers needed to expand their transportation network. Turnpikes, while preferable to unmaintained roads, were hardly the ideal solution to the region's problems, as an increase in coal traffic was made difficult by a number of factors. Many turnpikes could not reduce their tolls in order to attract more of the carrying trade as a condition of their corporate charter. For example, the Manchester Turnpike Company was forbidden by the legislature to reduce tolls so that the annual profit netted less than 15 percent of expenditures. Such provisions were intended to protect investors in the roads, but they did so at the expense of carriers. This practice was common to Virginia turnpike charters—indeed, nearly every American transportation company's charter—in the early nineteenth century. The Little River Turnpike (1802) had a 15 percent minimum on profits, and the Allegany Turnpike (1800) had the authority to charge double tolls until its line was completed.[49] The Richmond Turnpike Company, which linked the Deep Run coal pits to the city of Richmond, also had a 15 percent minimum on toll profits and explicitly prohibited coal wagons from making the return trip to the mines toll-free.[50]

As coal traffic swelled during the War of 1812, furthermore, turnpikes found that the higher toll revenue hardly covered the increased wear and tear on their roads. In response, many turnpikes asked the legislature for the right to charge higher tolls and for the power to limit the amount of coal that wagons could carry. Legislators often granted this request, much to the chagrin of local colliers. "I understand that most of the Gentlemen who own coal pitts or run waggons on the turn pike road, are very much displeased with those Gentlemen in the legislature that voted for the law regulating the turn pike road," William Pope, a delegate from Powhatan County, wrote to Harry Heth in 1812. Turnpikes became expensive to operate as a result of the increased traffic in coal, which further alienated the coal trade and drove transportation costs even higher. The Richmond Turnpike Company complained in 1813 that narrow-wheeled wagons filled with more than one hundred bushels of coal created deep ruts in its road which required constant repair and asked for a charter amendment allowing the company to increase the tolls on wagons carrying more than sixty bushels. Although the legislature continued to pass a great number of turnpike charters, "whose very names would fatigue the reader to peruse," in the words of one 1816 observer, these roads proved of limited use in expanding the Richmond basin's coal trade. At the time ton-mile rates for wagon transportation averaged about thirty to seventy cents. Since the price of coal averaged about three or four dollars a ton, overland transportation costs quickly bit into potential profits.[51]

Drawing upon the established pattern of transport in the British coal trade, colliers in the Richmond basin turned again to the James River Company's water navigation. Unfortunately, coal transport along the James River involved a number of costly and damaging transfers even after expansions were made following the War of 1812. The lack of any integrated or centralized system of coal transportation along the James, moreover, reduced the quality of the coal. After stockpiling coal along the banks of the James, carriers often shoveled the coal into wheelbarrows and deposited it into barges. Upon reaching the water basin at Richmond, the coal was again shoveled into wheelbarrows, carried to Richmond's tidewater port at Rockets, and dumped into warehouses. The coal then was shoveled onto wheelbarrows yet another time and loaded into the hold of the coastwise vessel to carry it to market. Without a staith to direct the flow of coal from one carrier to another, the rough handling often broke Virginia bituminous into smaller and smaller chunks as it was transferred, which reduced the combustibility of the mineral and lowered its value. If the coal came from the south side of the James River in Chesterfield County, where the James had no regular navigation, the coal was bounced around on wagons, which similarly pulverized it into a less useful commodity.

As a result, urban retailers found the coal shipped from Richmond less than satisfactory even when they asked for more to sell. For example, J. P. Pleasants of Baltimore complained to Harry Heth in 1811 that "the quality of the Coal lately recd had been so indifferent that I have lost the sale of a great

deal. It is not one, two or three who complain, but all." Thomas B. Main of
Boston wrote to Heth in December 1815 that his last shipment was "small and
dirty" and must have been the "scrapings of the yard." Main decided to can-
cel any orders for the next spring and summer, and Heth lost a valuable contact. Chesterfield County colliers complained to the legislature as late as 1824
that the existing system delayed and damaged their shipments as they worked
through the various locks and transfers. "The Quality of the coal is so naturally injured," miners argued, "that it can never gain a sufficient character in
the northern markets to offer an inducement to us to use [the canal]." In the
minds of Richmond basin colliers, these problems wiped out any cost savings
of shipping coal via the James in the years following the War of 1812.[52]

As problems with access to the canal emerged as a major criticism of the
James River Company, the company's failure to accommodate the coal trade
adequately became a familiar complaint to the legislature, especially among
mining interests along the route. In 1818 a group of colliers from Chesterfield
labeled the James River Company an "odious monopoly, by affording the
Planter, the Farmer, and Merchant of the upper country, a choice of markets."
At the same time that the coal trade suffered, however, traffic in tobacco,
wheat, corn, and other products of nearby plantations increased along the
James route. The Chesterfield colliers submitted testimony that confirmed the
problems of carrying coal on the James River canal. According to one carrier,
coal barges should hold 180 to 240 bushels each, but the low level of water in
the James only permitted carrying sixty or seventy bushels at a time. Veteran
collier Orris Paine testified that boats carried less than 70 percent of their capacity on average.[53]

In 1820 dissatisfaction with the progress of improvements along the James
spurred the legislature to reorganize the James River Company and to authorize the Board of Public Works to purchase two-fifths of the company's
stock—a move that made the Commonwealth of Virginia the controlling interest in the firm. The charter amendment that placed the James under state
control included a provision to enlarge the navigation of the James to allow
boats carrying a thousand bushels of coal to pass from Pleasant's Island to the
tidewater.[54] Throughout the 1820s and 1830s, however, improvements made
on the James never satisfied the nearby owners of coal lands, who remained

frustrated with the improved river and canal as a reliable outlet to market. In 1834 a group of colliers complained that their business had "not received that encouragement that those engaged in it had reason to expect, and the legislature intended to give." If the James River Company did not improve their amenities for carrying coal, they warned, "their businesses in the future must be carried on under great disadvantages, if not wholly abandoned."[55]

The location of the James River Company's canal and improved navigational route formed another complaint. The company's waterway intersected the Richmond basin on the north side of the James, therefore abandoning the rich mines of Chesterfield County. Mining interests found the lack of access for colliers on the south side troubling. In the winter of 1824 Chesterfield colliers often complained to the legislature about access to the north-side canal. Throughout the 1820s and 1830s colliers on the south side attempted to get some form of relief, whether from reduced tolls on the Manchester Turnpike, by forcing the state-run James River Company to provide access to the south side of the river, or by the construction of an entirely new canal on the south bank. In nearly every instance, however, their pleas were ignored. Meanwhile, in 1832 the legislature authorized the Manchester Turnpike to double its toll rates on coal.[56]

Other canal companies tried to service the coal trade but ran into similar problems. The Tuckahoe Canal Company, incorporated in 1827, linked many of the coalfields on the north bank to the James River system and was completed by December 1830. Coal miners in Henrico County, however, found the Tuckahoe company's improvements inadequate for coal transportation. In 1835 they complained to the legislature that the company's toll structure and improvement strategy were "very keen sighted to their own interest, but blind to and utterly regardless of every other consideration," especially that of the coal trade.[57] As with the early turnpike companies in the region, improvement companies servicing the Richmond basin appeared interested in regaining their stockholders' investments as quickly as possible rather than developing steady traffic along their lines over the long term. This strategy helped neither party in Henrico County, and consequently the Tuckahoe Canal was completely abandoned by 1840.[58]

Despite the high hopes of many Virginia colliers, the James never evolved into an efficient means of transport for the coal trade. From the perspective of the James River Company this made sense, as the majority of toll revenue (and subsequent dividends) came from agricultural products. In 1827, for example, the company collected $41,279.87 in tolls on their descending navigation. Tobacco produced 60 percent of the revenue, and agricultural commodities such as wheat, flour, and corn provided 22 percent. Coal was a significant source of revenue, amounting to 13 percent of toll revenues, but it hardly dominated the James River Company's interests. Two years later, despite various improvements along the route, the company still favored plantation products, as tobacco produced 42 percent of the tolls, other agricultural products 38 percent, and coal 16 percent. As these figures suggest, the company was willing to carry coal, but the products of the plantations upriver from the coal mines dominated traffic on the James River.[59] Until the 1820s, moreover, the tolls on the James River Canal were not based on the distance traveled and, therefore, acted as a subsidy for the farmers and plantation owners farther up the river. The company eventually stopped the practice of paying one flat toll rate for carriage, but by then the flow of traffic had been well established—a flow that largely neglected the needs of the coal trade. The toll structure of the James River Company and the lack of a water link on the south bank complicated the expansion of the Virginia coal trade.[60]

The reliance on slave labor meant that Richmond basin colliers encountered problems in the postwar decades. Heth employed a number of his own slaves in his Black Heath mines, but he usually needed an additional fifty to one hundred additional workers. If he could not hire enough slaves from surrounding plantations, he scrambled to fill out his labor force with free black and white unskilled laborers from Richmond. The inherent hazards of coal mining—fires, shaft collapses, and floods—made many owners reluctant to hire out their slaves for such dangerous work. One planter in 1812 refused to replace a slave who had fled Heth's coal mines and returned home. "I have been trying to hire hands for you ever since I saw you & have not as yet procured a single one," one friend wrote to Heth in 1819, "& I am afraid I shant be able to get one single one in the neighborhood." The high cost and scarcity of slave labor remained a particular expense of raising coal in the Richmond basin throughout the early nineteenth century.[61]

In addition to raising costs, the use of slave labor inhibited the application of new mining technology in the Richmond basin. In the years prior to the War of 1812, Harry Heth attempted to contract with local engineers to construct steam engines for pumping water from his mines. These efforts proved ineffectual, but blame for these failures cannot be laid directly at the feet of the slave labor economy.[62] Three years later Heth contacted Oliver Evans, the noted steam engine manufacturer of Philadelphia, about sending someone to design and implement a steam-powered system for pumping water from his mines. Evans agreed to build him an engine but refused to send any of his employees to Virginia because, he explained, "the workmen here have embraced a prejudicial idea of the customs of your country, they think that if a master mechanic goes into your employ, and will refuse to work a task himself but keep himself clean and talk big and help himself freely to brandy, wine &c, that you will treat him as a gentleman." "But," Evans admonished, "if he lays his own hands vigorously to the work (which he will be compelled to do because slaves cannot do it) you immediately rank him with slaves, with whom he is forced to work, and will not think him entitled to more than half the wages of a gentleman mechanic, who does not earn a cent by his hand." The steam engine manufacturer also balked at the idea that slave labor could operate his steam engine. "I fear [you] have wrong ideas if you think slaves can keep a steam engine in order," he wrote. "A man must be free before his mind will expand so much." Apparently, the pumping engine of Evans's design failed to yield results. "If you had performed with it not only you but all the pits about Richmond might have been receiving the benefits of my invention," Evans lamented in 1815, "and I might have had the profits of making the Engine."[63]

The delay in implementing steam pump technology to solve mine drainage problems exacerbated some endemic problems of industrial slavery. Miners referred to activities that did not directly result in the raising of coal as "dead work." As shafts in the Richmond basin bored deeper into the earth, the definition of "dead work" expanded to include propping roof supports along the mine shaft, ensuring proper ventilation in the mines, and, most important clearing the mines of water. Particularly waterlogged mines resulted in hundreds of lost hours for Virginia colliers. In later decades dead work was the responsibility of the miner, who was paid for the amount of coal raised and not by the hour. But for Virginia colliers, who hired labor at a fixed rate, dead work immediately cut into potential profits. "At the Christmas holidays when the hired negroes went home and before we hired others, the whole of those workings were filled with water," one miner reported. "It required constant work for ten days to get it out. Afterwards we could keep it under in two hours every day or less." Thus, even if colliers used skilled slave labor in their mines, they lost money whenever those workers raised more water than coal.[64]

Without a local, or even national, technological community to rely upon, Virginia colliers sought water removal solutions from overseas. In 1814 the state of war between the United States and Great Britain could not keep Heth from writing to a friend in London in order to seek a British engineer for his mines. Eventually, he did employ two Scottish miners and contracted the British firm of Boulton and Watt to construct a steam engine pumping system that finally worked. Although a manager at Black Heath predicted that the new steam engine would "afford a vast quantity of the most elegant coal," the charges for shipping machinery from distant manufactories and the high cost of training skilled workers to maintain steam engines made British technology expensive to implement in Richmond basin mines. Heth himself estimated that, although he had invested seven thousand dollars in steam engine technology by 1816, he was not confident that it would cut costs adequately. "The Engineers who are putting it up, appear sanguine," he wrote to a business contact. "I hope they may be right." Without an indigenous community of mechanics to draw upon, Virginia colliers found the application of effective technology to their mines to be rife with delays, unanticipated expenses, and other setbacks. Slavery exacerbated these problems by keeping the cost of labor high and thus frustrated attempts to integrate much-needed technological improvements in eastern Virginia's mines.[65]

Despite their problems securing a supply of labor, the high costs involved in hiring and purchasing slaves, and technological difficulties, Heth and his colleagues in the Richmond basin never lacked opportunities to expand business. Heth's profits increased during the period after the War of 1812, but, as table 1.2 demonstrates, his mines could not match prewar production levels. Profit margins actually increased on a per-bushel basis, but transportation

Table 1.2. Harry Heth's Coal Business, 1810–1817

	Bushels Shipped	Profit (Real $)	Profit (1810$)	Profit per Bushel (1810$)
1810		$52,092.34	$52,092.34	
1811	704,307	$43,647.85	$40,885.33	5.8¢
1812	830,742	$50,718.80	$46,914.89	5.6¢
1813	125,488	$8,155.42	$6,286.47	5.0¢
1814	45,867	$0	$0	0
1815	446,828	$68,778.73	$55,022.98	12.3¢
1816	503,005	$75,592.77	$66,199.59	13.2¢
1817	567,245	$78,540.89	$72,650.32	12.8¢

Source: Heth Papers, UVA; McCusker, *How Much Is That in Real Money? A Historical Price Index for Use as a Deflator of Money Values in the Economy of the United States.*
  Note: These figures are based upon Heth's own calculations, which run from 1810 to 1817.

and the burden of slave labor thwarted attempts to increase production. Although this suggests that Heth could still make money mining coal in the postwar market, his inability to adopt new technologies and increase the output of high-quality coal threatened the future profitability of Black Heath coal. Once the pent-up demand for coal caused by the embargo and War of 1812 subsided, the Virginia coastwise trade was vulnerable. Later reports suggested that there was plenty of coal remaining in the Richmond basin, but, as the Old Dominion's coal trade came into direct competition with anthracite coal in the 1820s, the poor quality and relatively high price of Virginia coal made it ripe for substitution.[66]

During the early republic many writers attributed to coal the potential for national economic self-sufficiency. To contemporary observers the bituminous region of eastern Virginia held the nation's best hopes of cultivating a dynamic American coal trade and therefore drew the attention and praise of early political economists. If Richmond could not be the new republic's Newcastle, it could at least serve the growing need for mineral fuel for the immediate future. From the very beginning, however, the Old Dominion's colliers struggled to increase production. Raising coal amid plantations, as it turned out, did not serve the future of the trade very well. Many of the short-term problems arose as colliers in eastern Virginia struggled with labor shortages and other problems associated with industrial slavery. Despite national aspirations, the nation's first coal trade ran into local limitations.

The presence of slave labor, however, was less significant to the long-term potential of the Richmond basin, as miners there raised quite a bit of coal with both free and unfree labor. The impact of Virginia's conservative and agrarian political economy—manifested in the intransigence of the James River Company, the indifference of local legislators, and the lack of state-level support for a strategically placed industry—inflicted much more damage on the future prospects of the Richmond basin in national markets. Colliers in eastern Virginia required a sustained effort to underwrite the coal trade's transportation network by the James River Company in its various private and public forms. They also needed a local technological community that could address the myriad challenges that mining presented to them. Slaves could mine coal well enough, but, as the wider impact of slavery helped calcify Virginia's conservative political and economic order, the Richmond basin's coal trade failed to exploit its advantageous position in U.S. markets.

Slavery was not the main problem facing the Richmond basin, as a pattern in the Old Dominion emerged during these formative years of the American coal trade. As colliers labored in a state structure devoted to the preservation of local interests, they often found themselves banished to the periphery of the political economy. State government officials, the board of the James River Company, and the Virginians who operated plantations in close proximity to the Richmond basin demonstrated little animosity toward their neighboring colliers. But they did reinforce a political and economic system that frustrated rapid and dynamic change by preserving established agricultural interests. Since colliers in eastern Virginia could not lay claim to a long tradition of privilege—even those who owned and ran plantations such as Harry Heth— the atomized nature of politics in the Old Dominion worked against their interests. In the end the institutional arrangement of Virginia's coal trade left its colliers dependent upon a network of northern merchants and wholesalers. Even the most ambitious collier was bereft of state support in transportation, reliant upon an unstable and expensive labor force, and unable to draw upon indigenous technological communities. The Richmond coal basin's dependence upon outside actors made it particularly vulnerable to substitution.

To the north of the Richmond basin another alternative to wood or foreign coal rose during the decade following the War of 1812. Pennsylvania an-thracite, or stone coal, was known to be plentiful, but it appeared to be of lim-ited utility and accessibility during the late eighteenth and early nineteenth centuries. Local boosters might make vigorous claims for it, but conventional wisdom literally reduced their claims to shouts in the wilderness. In a few short decades, however, these reputations had completely reversed with the emergence of anthracite coal in Pennsylvania as an important domestic and industrial fuel—a process that owed a great deal to the actions of Pennsylva-nia's energetic political economy and Virginia's sluggish response to develop-mental opportunities. Coal remained a political mineral of national impor-tance, but, as the rise of anthracite coal suggests, state governments assumed the burden of developing this valuable commodity in distinctive ways.

The Commonwealth's Fuel

The Rise of Pennsylvania Anthracite

*O*n a pleasant July evening in 1821 William H. Keating, a mineralogist and chemist, delivered a lecture on the state of mining in the United States before an audience of scholars, scientists, and prominent citizens at the American Philosophical Society in Philadelphia. The excavation of minerals, Keating argued, stood as one of the "the noblest and most interesting pursuits to which the attention of man can be called." Mining served as the harbinger of European civilization, and each nation developed its own style. The French, Keating maintained, developed a first-rate school of mines, while the ingenuity of German miners in culling the scarce mineral resources of their country was first-rate. Yet the highest praise within Keating's depiction of national systems of mining was reserved for the British. It was in Great Britain, he argued, that miners elevated their trade to the "highest degree of perfection" and helped cultivate the nation's natural stock of coal and iron for maximum economic benefit.[1]

When Keating turned his attention to the mining pursuits of his own nation, however, his tone and demeanor took a less sanguine tone. Although some small diggings appeared here and there, Americans appeared only mildly interested in mineral extraction. "Upon the whole, we think we may be warranted in saying, that there are as yet no mines in activity in the United States," Keating argued, "and we may consider the undertaking of mines on a regular system as a new branch in this country." The vast mineral wealth of the United States only exacerbated the failure of Keating's fellow citizens to recognize the value of mines to economic growth and independence. "Fuel is abundant and cheap in this country," he maintained, "and it may easily be brought to the places where it is needed, on account of the great facilities afforded by internal navigation." But without some effort on the part of Americans to cultivate mining, he concluded, the republic's many blessings in mineral deposits would remain obscured by ignorance and indolence.[2]

Keating's definition of a vigorous mining region, in which he invested both personally and professionally, eventually found expression in eastern Pennsylvania at about the same time that the high hopes of Virginia's colliers sunk. It is tempting to see the rise of anthracite coal as a matter of simple economic substitution in which the knowledge of anthracite's strong and efficient flame allowed urban consumers to switch immediately from wood and bituminous coal to Pennsylvania's stone coal. Perhaps some conservative holdouts relied upon other kinds of fuel, but, as rational economic actors, the residents of Boston, Philadelphia, and New York could not help but make the change to anthracite. Prior historians of the coal trade certainly made this assumption; their story of anthracite coal's rise in the early nineteenth century championed individual entrepreneurs, consumers, and forward-thinking canal companies. Once people understood that anthracite coal requires a kindling fire to ignite and needs to be kept separate from its ashes while burning, any cost-benefit analysis would favor the efficiency of stone coal over the more traditional fuels. Why, after all, would someone continue to use Virginia coal when the evidence suggests that anthracite coal burned longer, cheaper, and brighter?[3]

At a theoretical level this explanation makes sense. But some economists and historians of technology suggest that, even in the face of increasing returns, inefficient technologies can be locked in and continue to dominate more advanced methods. Often some combination of historical events and evidence of increasing returns is necessary to overcome the initial advantage of a technological process, even if it is less efficient. Although anthracite coal's flame burned hotter and longer, their models suggest, we cannot assume that consumers immediately switched to this new fuel technology. To put it more directly, when a blacksmith in New York in 1807 tossed a chunk of anthracite into his furnace, the stone coal was unlikely to ignite. If a middle-class family in Philadelphia dumped a load of anthracite in their hearth during the lean years of the embargo, the dense mineral was much more likely to squelch their fire. This initial impression is not easily dismissed, and it complicates assessments of the early American coal trade in terms of fuel efficiency and relative prices—something had to bridge the gap between the familiar practice of fuel use and the new technology of anthracite. Thus, the rise of the Pennsylvania anthracite coal trade must be assessed with particular attention to the technological and political context in which it occurred.[4]

In the struggle for the early coal trade a number of actors entered the story, complicating any simple narrative of technological substitution. Individual entrepreneurs and forward-looking canal companies were involved, of course. But political factors in Virginia and Pennsylvania also played a critical role in shaping the nature of this transformation in two major ways. First, as demonstrated in chapter 1, Richmond basin colliers found that their marginal position in the Old Dominion's plantation society undercut an opportunity to solidify their hold on eastern markets in the years following the War of 1812. As Virginia's political system reinforced the interests of slave-based agriculture, Richmond's coal trade struggled to increase production. The second political factor in the early coal trade involved a bevy of institutions that helped promote the cause of stone coal in Pennsylvania. Although early boosters of anthracite coal acted on their own accord and without public help, during the 1820s and 1830s more and more Pennsylvania-based institutions leaped into the fray. Anthracite coal materialized as the "commonwealth's fuel," and institutions such as the Franklin Institute and the Pennsylvania legislature blended the twin causes of state-level patriotism and economic development in order to promote its development. Without an institutional apparatus of their own to promote Virginia coal, colliers in the Richmond basin faded in importance.

By the 1830s the escalation in mining which Keating had so desired in 1821 was fully under way in the United States—a process that owed a great deal to the help of state-level institutions in Pennsylvania.

COAL AND THE KEYSTONE STATE

To the north of the Richmond basin, Pennsylvania's coal seams lay in a strikingly similar pattern as Virginia, for the Appalachian mountains divide Pennsylvania, too, into western and eastern sections. But the spine of the Alleghenies runs diagonally across the state, thus limiting Pennsylvania's eastern plain to the southeastern region of the state around Philadelphia. Pennsylvania can be divided into three major sections, each reflecting a stage of immigration of white Europeans to the area. Settlers of English descent and of the Quaker, Anglican, or Presbyterian faith were the first whites to arrive in large numbers in Pennsylvania. They populated the city of Philadelphia and its prosperous agricultural hinterland. Pennsylvania's midsection, a region of rich farmland hemmed in by the ridges of the Appalachian Mountains, was originally populated by German settlers who arrived in the mid-eighteenth century. Central Pennsylvania's landscape very much resembled that of Virginia's valley: rolling hills occupied by small, but thriving, family farms. Western Pennsylvania's mountainous terrain was first settled by Scotch-Irish Presbyterians in the second half of the eighteenth century. But at the onset of the nineteenth century most of western Pennsylvania remained sparsely populated, save the area surrounding the growing town of Pittsburgh.[5]

Pennsylvania's western counties, however, served as the location of its earliest coal industry. Western Pennsylvania, like Virginia, holds vast reserves of bituminous coal which underlie over fourteen thousand square miles of the state. Aside from the Broad Top region, a coalfield centered in Bedford, Fulton, and Huntingdon counties in the south-central portion of the state, the vast majority of bituminous coal seams lie in western Pennsylvania. The most prominent of these, the Pittsburgh Seam, spreads across an area of nearly six thousand square miles into the bordering states of Ohio, West Virginia, and Maryland and averages from five to eight feet in thickness. The military occupation of western Pennsylvania during the French and Indian War produced the first known references to the Pittsburgh Seam, as Colonel Hugh

Mercer, in command of troops in Pittsburgh, reported the nearby presence of "excellent coal and limestone."[6]

Pennsylvania's western metropolis, Pittsburgh, used bituminous coal for domestic and smithing purposes in the late eighteenth century. In 1784 Arthur Lee, a Virginian visiting the forks of the Ohio, noted, "The coal is burnt in the town and considered very good." "The great supply will be uncommonly advantageous in the future settlement of this region," a German visitor to Pittsburgh noted three years later; "coal is found not only here but in almost every hill on both sides of the Ohio throughout the western country and most of the mountain valleys contain coal-beds." Most of Pittsburgh's coal came from "Coal Hill," now known as Mount Washington, and was easily obtained by excavating exposed veins along the side of the mountain. As supplies of wood became less abundant, residents increasingly burned coal both at home and work. As the Pittsburgh region grew in population, so did its use of mineral fuel. In 1800 an English observer noted, upon approaching Pittsburgh, "A cloud of smoke hung over it in an exceedingly clear sky, recalling to me many choking recollections of London."[7]

Early coal mining in western Pennsylvania, like its counterpart in Virginia, served as a supplementary activity to agriculture. At the turn of the nineteenth century local farmers mined coal on their land during slack winter months. Since coal veins along the Monongahela River broke the surface, Pittsburgh-area miners hacked at the seams with picks and shovels and rolled bundles of coal down the slope of the riverbed into boats. These part-time miners then sold their product in Pittsburgh for about $1.25 a ton. Within a few decades this informal coal trade saw the appearance of full-time miners and brokers. Two of these early entrepreneurs, James and Robert Watson, made regular shipments from the Monongahela Valley mines to New Orleans. Throughout the years of the early republic, however, the bulk of Pittsburgh's coal trade remained small in scale and mostly served local markets.[8]

To the east Pennsylvania contained nearly all of the reserves of anthracite coal in North America. The four major fields of anthracite coal ran parallel to one another from southwest to northeast and were arranged in three distinct production areas about fifty miles long and five miles wide: the Schuylkill, which includes the western middle field and the southern field west of

Tamaqua; the Lehigh, consisting of the western middle field and the southern field west of Tamaqua; and, finally, the Wyoming, containing the northern anthracite field. Compared to most bituminous coalfields, the anthracite districts appeared quite tiny, as they were largely confined to Pennsylvania's Lackawanna, Luzerne, Schuylkill, and Northumberland counties, but these densely packed fields contained an estimated twenty-two billion tons of anthracite coal.[9]

Pennsylvania's anthracite coal region lay tucked away in its eastern mountains, where transportation to the seaboard was costly and difficult, if not impossible at times. The Richmond field was close to tidewater and, like the Newcastle and Firth of Forth coal regions in Great Britain, maintained easy access to sea transportation. Nature seemed to favor the Old Dominion's early coal trade, and at the time of the American Revolution these anthracite coal regions did not share the national notoriety of the Richmond bituminous field. Pennsylvania's eastern mountains remained rugged, unimproved territory whose white settlers scratched out a living tilling the mediocre soil and hunting in the vast forests that blanketed the area.

In 1791 one of these hunters in the Lehigh region discovered chunks of a black stone clinging to the roots of a tree that had been blown down in a storm. A year later the Lehigh Coal Mine Company mined a small amount of coal from the area but quickly folded due to the hardships encountered in shipping the coal down the unimproved Lehigh River.[10] Residents of the Schuylkill region found similar outcroppings of stone coal, but they regarded it as useless until 1795, when a local blacksmith burned it in his shop. The Wyoming Valley dates the use of anthracite back to Judge Obadiah Gore's burning stone coal in his blacksmith's fire at Wilkes-Barre in the late 1760s. By the first decade of the nineteenth century local residents intermittently substituted anthracite coal for wood in their hearths, yet systematic mining of stone coal would not blossom until the years following the War of 1812. A few members of Pennsylvania's budding scientific community noticed anthracite's enormous potential, but the vast majority of consumers considered stone coal an inferior product. Richmond or British coal thus remained the standard mineral fuel in large eastern cities during the first two decades of the nineteenth century.[11]

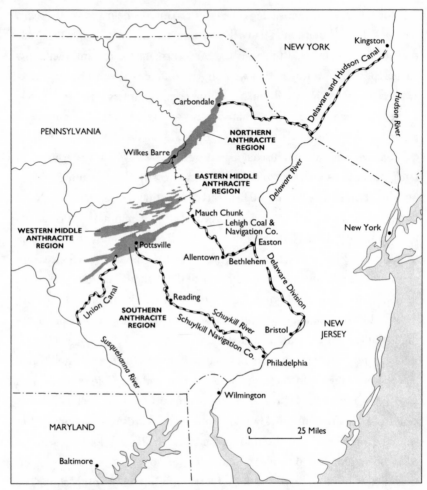

Map 3. Pennsylvania's anthracite region

The War of 1812 disrupted the nation's coal trade and revealed the eastern seaboard's dependence upon Richmond coal, but it also offered a unique opportunity for anthracite to gain ground in urban markets. The presence of the British navy in American harbors certainly made the regular shipment of Virginia coal difficult, and by 1813 the inability of Richmond basin producers such as Harry Heth to get enough Virginia coal to Philadelphia spawned a full-blown fuel shortage in many urban areas of the United States. Jacob Cist of Wilkes-Barre leapt at the chance to exploit this crisis by introducing Lehigh anthracite as a cheap alternative to bituminous coal. He shipped five hundred tons of Lehigh anthracite to Philadelphia in the fall of 1814, which more than satisfied the city's demand for fuel through the winter months of 1815. News of impending peace with Great Britain, however, quickly halted the growth of Lehigh coal in Philadelphia. Blacksmiths there found anthracite an adequate substitute during the critical shortage, but the difficulties in burning stone coal and the return of Richmond basin coal with the end of the blockade again moved anthracite to the margins of the coal market.[12]

The war created a temporary disruption in the supply of Virginia coal, in other words, but did not result in its permanent displacement. In the large urban markets of Philadelphia, New York, and Baltimore, preferences for wood and bituminous coal fuels made it difficult for anthracite to make major inroads. When Virginia producers could not get enough of their product to market, most consumers used wood or foreign coal. Cist regularly bribed Philadelphia journeymen to use stone coal in their forges, but these paid endorsements did little to promote sales after the fuel crisis had passed. Henry Abbet, a master blacksmith and the director of Philadelphia's Mutual Assistance Coal Company, for example, was willing to use anthracite during the war but resumed purchasing Virginia coal in 1815. The reason for this reluctance to shift from bituminous coal was technological: anthracite was simply too difficult to ignite in conventional fireplaces and furnaces. The War of 1812 demonstrated that in a pinch urban consumers would buy stone coal, but under normal conditions Virginia coal remained the preferred domestic mineral fuel. In order to cultivate urban markets, the boosters of anthracite needed more than a temporary disruption in the supply of wood or bituminous coal; they required a sustained transformation of fuel technology.[13]

Encouraged by his brief success during the war, Jacob Cist of Wilkes-Barre spearheaded the efforts to change fuel preferences. Cist and his business partners, Isaac Chapman and Charles Miner, spent the fall of 1814 in Philadelphia promoting anthracite coal through personal visits, handbills, and newspaper advertisements. They sold some anthracite, but the most important result of this trip was the publication in April 1815 of two hundred copies of a pamphlet entitled *On the Importance and Necessity of Introducing Coal, Particularly the Species Known by the Name of Lehigh Coal into Immediate and General Use.*

Cist described anthracite's immense power once ignited and sought to dispel the conventional wisdom concerning the difficulties in burning stone coal. The pamphlet also noted that, because anthracite produced no soot, the domestic consumer need not worry about his or her chimney or stovepipe catching fire; "neither will the misery of a smoky chimney ever be endured where this fuel is used."[14]

Cist included a number of testimonials in his pamphlet, all confirming the superiority of anthracite over wood, charcoal, and bituminous coal. Frederick Graff of the Pennsylvania Bank, for example, noted that Lehigh coal was not only less expensive than wood but also burned cleanly, with no smoky or sulfurous smell, "which is a great objection to all other coal for family use." Other testimonials compared anthracite with Richmond coal explicitly. David Hess, a smith and gun barrel maker of Northampton, Pennsylvania, estimated that a single bushel of Lehigh coal "is equal in durability and value to nearly three of Virginia." Joseph Smith of Bucks County estimated Lehigh coal to be worth twice as much as Richmond coal; he surmised that one bushel of anthracite equaled ten to twelve bushels of the best charcoal. "These coals will no doubt prove the cheapest, most durable, cleanly, and pleasant fuel for warming apartments," declared the inventor and steam engine builder Oliver Evans, "as well as for other useful purposes." Cist's efforts seemed to pay off. Anthracite sales increased from 365 tons in 1820 to 33,393 tons by 1825.[15]

Two other studies boosted anthracite's superiority as a domestic fuel and furthered the acceptance of stone coal in urban markets, although from a scientific perspective. The first was done by Marcus Bull, the president of the North American Coal Company in the Schuylkill region, whom contemporaries described as a "gentleman, well educated, of a philosophical turn, prepossessing appearance and a great suavity of manners." Bull attempted to test the relative quality of heat produced from various types of wood and coal. In November 1823 he burned equal weights of woods such as ash, birch, maple, oak, and pine and Lehigh and Schuylkill anthracite, Newcastle and Richmond bituminous, and charcoal in a twelve-inch-high stove situated within an eight-foot square compartment. Bull noted each material's specific gravity and measured the time that one pound of fuel could maintain a temperature ten degrees hotter in the compartment than in the exterior room. He took care to follow scientific procedures in his experiments and consulted several experts on his methods, including University of Pennsylvania professor of chemistry Robert Hare.[16]

His experiments aimed at "practical utility, rather than scientific research," so Bull formulated a table to compare the value of different fuels. His numbers demonstrated that anthracite coal was the most efficient fuel for heating the interior of homes. "The value of a given quantity of fuel," Bull argued, "is directly proportional to the *time* that a given weight of it maintained the air of the room at a given temperature, and also to its *weight*."[17] Thus, the basis for Bull's comparison was the amount of time that identical quantities of each material could maintain the 10 degree difference between the interior and exterior rooms. Most woods did so for about six and a half hours (table 2.1). American and foreign bituminous coal maintained the 10 degree standard for a little over nine hours. Anthracite coal, however, sustained the desired temperature difference for at least thirteen hours. Bull's conclusions inflamed the imagination of many Pennsylvanians, especially those directly involved in the anthracite trade.[18]

Yet the commercial aspirations of Bull's work compromised the impact of his experiments within the American scientific community. In 1826 Bull applied for the Rumford Premium, an award sponsored by Boston's American Academy of Arts and Sciences, for the best discovery or improvement relating to heat or light. He quickly ran into major opposition. The committee in charge of awarding the premium cited major flaws with Bull's double room design, its inability to control the effects of outdoor atmospheric conditions,

Table 2.1. Comparative Values from Bull's Experiment

Fuel Used (One Pound)	10 Degrees Maintained
White Ash	6 hours, 40 minutes
Shell-Bark Hickory	6 hours, 40 minutes
Jersey Pine	6 hours, 40 minutes
Newcastle Coal (bituminous)	9 hours, 10 minutes
Richmond Coal (bituminous)	9 hours, 20 minutes
Lehigh Coal (anthracite)	13 hours, 10 minutes
Schuylkill Coal (anthracite)	13 hours, 40 minutes

Source: Bull, "Experiments to Determine the Comparative Qualities of Heat," 280–81.

and his use of dried wood as a fuel. Most people burn wood that is not completely dry, and, when this moisture vaporizes, the wood loses some heat, the
committee argued. Therefore, anyone who should "govern his practice" by
Bull's findings "would act under a perpetual error."[19] Two years later Jacob
Bigelow, Bull's primary critic, complained that "no great practical good is
known to have grown out of Mr. Bull's experiments." Bull immediately responded by citing the opinions of various consumers of anthracite and bitterly concluded that the academy would probably never respect his experiments. "What will they say of those statements which so forcibly prove the
correctness of my results? Will they try them and condemn them by their old
standard of *theoretical* accuracy," Bull asked of his critics, or will they claim
that "these are the mere 'opinions of the practical part of mankind, and *particularly of the manufacturing class,*' on whose results *no reliance* can be
placed?"[20]

A less controversial, but no less enthusiastic, endorsement of anthracite
came from the pen of Benjamin Silliman, one of the nation's most eminent
scientists. This professor of chemistry at Yale College in New Haven founded
the *American Journal of Science* in 1818 and hoped that the publication would
raise American science and technology out of its provincial condition. In 1822
Silliman's journal republished Cist's promotional pamphlet without any critical commentary. Four years later it published an article in Silliman's name entitled "Anthracite Coal of Pennsylvania, &c. Remarks upon Its Properties and
Economical Uses." Silliman's interest in anthracite is not surprising given that
he, like Marcus Bull, was a member of Philadelphia's American Philosophical
Society and was closely associated with Philadelphia's scientific community.
After all, the aim of the *American Journal of Science* was, in the words of Silliman's biographer, "in no small degree to nourish enlarged patriotism," and
what could be more patriotic and scientific than exploring a potential resource for American hearths?[21]

Silliman's 1826 article synthesized Bull's optimistic scientific approach
with Cist's promotional aims. He began his experiments by inserting specimens of anthracite and bituminous coal into wrought iron tubes and then
exposing them to extreme heat in a large furnace. The tubes were connected
with a flexible tube to a "hydro-pneumatic" cistern so that Silliman could ob-

serve the gas that each sample emitted when heated. He then burned each gas and concluded that "the comparatively abundant flame of the Pennsylvania anthracite must fit this fuel for some important purposes." For domestic use he found them to be comparable: "both the Lehigh and Schuylkill coal are applied with great advantage, both in parlour grates and in close stoves, for warming apartments, as well as in cooking." After giving detailed instructions on how to burn anthracite at home, Silliman remarked upon the mineral's abundance and listed eleven principle advantages of anthracite as a domestic fuel. Among them were its cheapness, its safety, the quality of anthracite's heat, and the fact that the burning of anthracite produced few foul by-products.[22]

As a man of science, Benjamin Silliman disavowed the outright promotion of anthracite as a domestic fuel, and he recognized that conventional hearths, furnaces, and stoves burned Richmond and British coal more easily. "It is not the object of the preceding remarks to depreciate the bituminous coals," he argued. "They are of great value and they certainly afford a heat *more at command* than the anthracites." Any doubt about Silliman's vigorous endorsement of anthracite, however, was dispelled by the nineteen testimonials championing anthracite which were appended to his article. Later that year Silliman further demonstrated his support for anthracite by writing to the American Academy of Arts and Sciences to endorse Marcus Bull's quest for the Rumford Premium. Subsequent articles in the *American Journal of Science* helped to keep anthracite coal in the intellectual spotlight.[23]

Despite the glowing results and publicity surrounding these experiments, Bull himself realized that anthracite use required special attention in the home. Scientists could demonstrate that anthracite burned hotter and longer than wood or bituminous coal under experimental conditions, but its practical use in fuel markets was another matter. "The difficulty of consuming small quantities of anthracite coal, in open grates," he wrote, "must operate to prevent its general introduction into use, unless this difficulty can be removed." Anthracite still required a high temperature to ignite. Once it was lit, moreover, it needed to be separated from the ashes it produced so that its unused surface continued to be exposed. Many fireplaces, stoves, and forges in the early 1820s—most of which burned wood or coal on an open grate—simply could not meet these requirements. The editors of the *Miners' Journal* gloried in the results of Bull's experiments, but they cautiously noted that "the question . . . can only be exactly determined by experiments having in view practical *convenience* as well as economy." Although Bull and Silliman demonstrated an innate superiority of anthracite as a fuel in the laboratory, the practical problem of how to get people to use it remained the primary challenge for boosters of anthracite.[24]

One institution that eagerly tackled this problem was the Franklin Institute for the Promotion of the Mechanical Arts, which was founded in Philadelphia in 1824. The Franklin Institute's mission was to encourage a more democratic propagation of technical and scientific knowledge while enriching Philadelphia's local economy and developing Pennsylvania's natural resources. Yet, along with the idea of an equitable distribution of technological knowledge, the founders of the Franklin Institute also sought to promote the industries of Pennsylvania. Thus, although it fashioned itself a nonpolitical, educational institution, the Franklin Institute's Pennsylvania-focused agenda gave both energy and clarity to the loosely organized campaign to promote anthracite in the 1820s and 1830s. It also provided a vital link between the dispersed individual efforts of entrepreneurs such as Jacob Cist and the coordinated efforts of large institutional actors.[25]

The institute accomplished this by sponsoring a series of industrial exhibitions from 1824 to 1838 and awarding gold, silver, and bronze medals for technological innovations. The highest accolade, the gold medal, was explicitly reserved for improvements in the utilization of Pennsylvania's coal and iron resources. The institute's *Franklin Journal and American Mechanics' Magazine* also made note of patents relating to the use of anthracite. Over a three-year period beginning in 1830, the journal reported on at least eight major innovations in the use of anthracite in the home. These inventions included cooking stoves, stoves for drying tobacco, and fireplace grates for burning anthracite.[26] The journal also consistently promoted anthracite by publishing favorable letters regarding its use. On one occasion the editors concluded, "We think our stone coal is as much to be preferred to the bituminous kind, in all the operations of the kitchen, as our servants now think wood, is preferable to anthracite." These articles highlighted innovations in grates, stoves, and fireplaces and often included explicit instructions on how to burn anthracite in the home along with a more technical description of the apparatus.[27]

The campaign for anthracite gained momentum in the late 1820s and early 1830s, as articles promoting both the superior efficiency and general utility of anthracite appeared in a number of urban presses. Newspapers played a critical role in dispelling the common assumptions about stone coal. It was a common reaction of the uninitiated, for example, to blow air on anthracite because it was slow to light, but doing so tended to restrict, not encourage, combustion. Articles and testimonials, often paid for by anthracite dealers, helped dispel these and other myths. They also translated the technological innovations in grates, stoves, and chimneys into a language that most residents understood: saving money. Urban residents at first wanted to dump anthracite into an open fireplace, as conventional wisdom dictated, but they eventually learned that a hearty flame would only spring up if they used grates to separate anthracite from its ashes. "It was necessary to combat and remove old and long established prejudices; and to satisfy the public that a saving would be made," the Schuylkill County Coal Mining Association reported in 1833. "Before coal could be of use, grates must be substituted for the open fireplaces, at a heavy expense, which was thought too great to be hazarded for an uncertainty." As boosters of anthracite eventually discovered, this process required the enlistment of larger forces in the Keystone State.[28]

AN UNCOMMON COMMONWEALTH

What kind of polity provided the framework for the anthracite coal trade? Pennsylvania's turbulent and complex political history during the revolutionary and early national eras laid the groundwork for the state's active support of economic growth and provided a striking contrast to the conservatism that permeated Virginia politics. Whereas the Old Dominion sought to preserve the interests of an established elite centered in the eastern part of the state, Pennsylvania's state government facilitated a constant struggle to balance power between East and West, city and country, farmer and manufacturer, and a host of other conflicts. Anthracite interests thus constituted only one cause among many, but in the end they found the Keystone State's polity a receptive atmosphere for their growth. So long as they could tie the expansion of their trade to the general prosperity of the state, the boosters of anthracite found succor among state officials when they enlisted the aid of government in the cause of stone coal.

The ability of state government to nurture the anthracite trade along with other economic endeavors can be traced back to Pennsylvania's revolutionary origins. As the Provincial Assembly of 1775 and 1776 wrestled with the question of severing Pennsylvania's ties with Great Britain, it also gave birth to a dramatic restructuring of Pennsylvania's governing structure. A "radical" coalition of Philadelphia intellectuals, the city's growing artisanal class, and western politicians led the charge to form a new state government that would offer greater influence to their particular interests. They immediately ran into significant loyalist opposition in the Provincial Assembly. The radicals called a Provincial Conference in May 1776 in response to a resolution from the Continental Congress, which was meeting in Philadelphia at that time. The Provincial Conference wrested control of the state from the Provincial Assembly, organized the state's military mobilization against British rule, and called for the election of a state constitutional convention later that summer.[29]

The resulting constitution of 1776, unlike Virginia's of the same year, delineated a departure from Pennsylvania's colonial past and erected one of the most striking state structures of revolutionary America. The convention created a unicameral legislature called the Assembly, a weak executive body called the Supreme Executive Council, and a troubleshooting committee known as the Council of Censors, which met every seven years to review the performance of the government. Legislative supremacy in Pennsylvania resembled the system developing in the Old Dominion. But, unlike in Virginia, all taxpaying men in Pennsylvania had the right to vote for representatives in the Assembly, the Supreme Executive Council, and the Council of Censors—a change from the colonial system, which enfranchised only free men owning fifty pounds worth of property.[30] The convention delegates very consciously empowered the newly expanded electorate by making all branches of government accountable in popular elections, by apportioning the Assembly in proportion to the number of taxable inhabitants, and by providing for the periodic reapportionment of the Assembly as Pennsylvania's population changed.[31]

Pennsylvania's new government soon became the focal point of opposition by "Anticonstitutionalists," or "Republicans," who attempted to undermine the radical state structure through an active editorial campaign in the 1770s and 1780s. Anticonstitutionalists feared that the omnipotent Assembly would be susceptible to factionalism and believed that its domination of a weak executive branch created a tyranny within the legislature in Pennsylvania. The Executive Council appointed both local justices of the peace and Supreme Court justices, critics noted, and any member of the judiciary could be removed by the Assembly for "misbehavior." Thus, Pennsylvania's Assembly controlled the judicial branch as well. The Anticonstitutionalists proposed altering the constitution to create a bicameral legislature, a stronger executive, and an independent judiciary. This, they believed, would remedy the evils of factionalism and limit the power of the legislature. Rather than having a state subject to the demands of a fickle legislative electorate, conservative Pennsylvanians championed a structure composed of three equally balanced branches of government.[32]

Despite efforts by Pennsylvania's radicals to preserve their 1776 constitution, conservatives won control of the Assembly in 1787, forced a state constitutional convention in late 1789, and ultimately secured the passage of a new constitution for Pennsylvania in 1790. Using the federal constitution as a guideline, the convention of 1789–90 created a system of more balanced government. The governor, elected by popular vote for up to two consecutive three-year terms, could veto legislation, grant pardons, and exercise control over the Pennsylvania militia. The 1790 constitution also abandoned the idea of a unicameral legislature in favor of a divided one consisting of a lower House of Representatives and an upper house, the Senate. The convention abolished the Council of Censors and made the judiciary more independent from legislative and executive authority.[33]

Although the constitution of 1790 represented a dramatic break from the radical state structure created in 1776, it nonetheless retained a few critical components of the radical years which would affect policy making in Pennsylvania for years to come. First, the 1790 constitution preserved the liberal suffrage requirements of its predecessor; any free man who had resided in Pennsylvania for two years and paid taxes upon his property could vote in local,

state, and federal elections. Second, both the House and Senate, like the unicameral Assembly, remained apportioned on the basis of population. The
1790 constitution, like its predecessor, moreover, made provisions for a periodic reapportionment (once every seven years) as Pennsylvania's population
changed. The retention of these three elements of Pennsylvania's 1776 constitution tempered the conservative nature of the state structure created in
1790.[34]

Political conditions in the 1790s immediately tested the ability of Pennsylvania's state-level institutions to alleviate sectional tension. Like the Cohees of
Virginia, Pennsylvanians west of the Alleghenies often viewed their neighbors
to the east with suspicion and outright hostility. A federal excise tax provided
the flashpoint for 1794's Whiskey Rebellion, and President George Washington's swift reprisal provided an immediate resolution to any armed resistance
to governmental authority in western Pennsylvania. Nevertheless, healing the
wounds between western and eastern Pennsylvanians in the long run required
state-level action. Over the next few decades, therefore, eastern delegates
found ways to mute the long-standing gripe that western interests were ignored in the legislature. Backcountry dissention in western Pennsylvania continued to fester throughout the early republic, but western delegates never
found themselves systematically marginalized from such vital areas as constitutional change or internal improvement policy as did their counterparts in
Virginia. The Pennsylvania legislature certainly witnessed its share of sectional tension in the early nineteenth century, but periodic reapportionment,
broad-based suffrage, and the distribution of turnpike and river improvement
projects tempered the divisive effects of the east-west divide in Pennsylvania
politics.[35]

In addition to potential feuds between western farmers and eastern urban-
ites, political blocs representing ethnocultural interests of English Anglicans
and Quakers, German Mennonites and Amish, and Scotch-Irish Presbyterians influenced Pennsylvania politics during the early republic. The commercial interests of Philadelphia, the needs of nascent manufacturers of iron, and
other economic factors added additional color to this mosaic of interests. The
infusion of national partisan issues during the Federalist-Republican battles
of the 1790s complicated matters even more. As ill-fated attempts to reform

the state's judicial and militia systems in the early nineteenth century suggested, diversity often hindered policy development in Pennsylvania. But Pennsylvania's political institutions proved malleable in dealing with these varied
perspectives. More often than not, delegates successfully formed broad-based
coalitions, swapped votes, and selectively employed party discipline enough
to overcome political paralysis.[36]

The relocation of the state capital provided a symbolic example of Pennsylvania's political flexibility in the face of economic, social, and cultural diversity. Philadelphia dominated the state's politics and served as the seat of the
legislature for the first twenty-three years of Pennsylvania's statehood. As the
state's population continued to grow in the areas west of Philadelphia, however, periodic reapportionment allowed for the state's political center also to
move westward. Delegates thus relocated the state capital westward to Lan-
caster in 1799. By 1812 western and central portions of Pennsylvania grew in
terms of population and influence and caused yet another move of the legislature, this time to Harrisburg, where the capitol remains to this day. Although
on the surface it seems a trivial matter, the physical relocation of Pennsylvania's capitol epitomized a willingness to bargain and accommodate among
disparate sectional interests.[37]

Anthracite colliers benefited from the Pennsylvania legislature's ability to
accommodate different economic interests throughout the industry's formation. The creation of the canals that linked anthracite mining regions to market centers bore the marks of state action from their earliest inception. The
first anthracite canal used the Schuylkill River to link the mines of the Southern Anthracite Field in Schuylkill County to Philadelphia. Ideas on how to improve the Schuylkill for navigation dated from the colonial era, but most plans
failed until 1815, when Philadelphia representatives ushered through a bill entitled "An Act to Authorize the Governour to Incorporate a Company to Make
a Lock Navigation on the River Schuylkill." It authorized the directors of the
Schuylkill Navigation Company (SNC) to raise $500,000 in initial capital in
order to improve the river over the 114 miles from Mill Creek in Schuylkill
County to Philadelphia. The act spelled out the physical specifications of locks
and dams along the navigation, and it set a maximum dividend of 9 percent
for the first five years of operation, with 15 percent as the upper limit after that

time. The charter did not authorize the company to own land beyond that required for the improvement, and an 1821 supplement to the firm's charter made this restriction explicit. In order to ensure the charter's passage through the legislature, supporters of the SNC accepted this property-holding limitation, which most likely originated from political opposition to the project. As other carriers serving the Philadelphia coal trade appeared later in the decade, the SNC's restriction on owning coal lands emerged as the defining characteristic of the Schuylkill anthracite region.[38]

Work on the Schuylkill navigation initially suffered from a winter-shortened construction season, a dearth of skilled engineers, and difficulties in raising adequate funds. In 1817 the Pennsylvania legislature aided the cause by purchasing $50,000 in SNC stock. Yet the state held a relatively small interest in the SNC during the firm's early years, as Philadelphia banker Stephen Girard alone subscribed to $60,000 worth of shares. By 1825 the navigation was open from Philadelphia to Mount Carbon, two and a half miles from of the chartered terminus. The opening of regular traffic along the entire navigation route in that year provided a much-needed boost to the prospects of the SNC, as the company had collected only $7,685 in tolls against the $950,000 in capital stock and $1,030,873.60 in loans accumulated by the firm. The founders of the Schuylkill navigation did not push the project with the coal trade in mind, but by the early 1820s its directors hoped that anthracite would play a large role in the future of the SNC. In 1821 Cadwalader Evans, the president of the SNC, predicted that the market in Philadelphia for Schuylkill coal would serve as a staple of the firm's revenues, instead of the agricultural and light manufacturing traffic that had barely sustained the company to that point. Evans estimated that Schuylkill anthracite paying a toll of six cents a bushel would "give the stockholders annually $120,000, and the city will have the vast advantage of purchasing fuel at one-half its present cost."[39]

Shipments of coal lagged during the early operation of the SNC, and commodities such as lumber, flour, and stone constituted the main traffic of the canal during this period. As suggested earlier, urban consumers harbored serious doubts about anthracite coal, which limited shipments in the 1820s. As the campaign led by Jacob Cist and the Franklin Institute progressed, coal traffic to Philadelphia increased substantially. By the early 1830s Evans's forecast for the SNC's coal trade proved true, and mineral shipments became the lifeblood of the navigation. As demonstrated in figure 2.1, coal traffic along the Schuylkill route rose to more than 100,000 tons by 1833, which constituted about 70 percent of both the total descending tonnage and toll revenues. In 1837 the SNC shipped more than 500,000 tons, which made up 80 percent of the descending tonnage and toll revenues.[40]

A second major carrier planned from the outset to develop the anthracite coal trade. The Lehigh Navigation Company used the Lehigh River to link the eastern portion of Pennsylvania's Middle Anthracite Field to the Delaware River and ultimately to Philadelphia. Josiah White, George Hauto, and Erskine Hazard formed the Lehigh Navigation Company in 1818 to improve the Lehigh River from Mauch Chunk (modern-day Jim Thorpe) to the Delaware River. Although the initial endeavor was to create a water navigation to Philadelphia, the coal trade loomed large in this endeavor. "I made diligent inquiry about the Coal on the Lehigh, & the C° which owned it to the amount of 10,000 acres," White reminisced years later, "and it occured to me that I

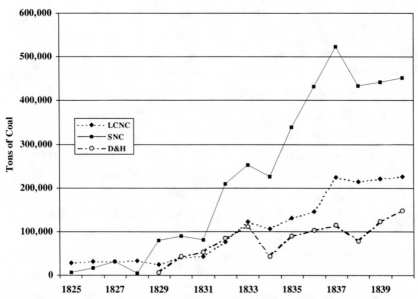

Fig. 2.1. Anthracite coal traffic on the three major tidewater canals, 1825–1840. (Annual reports of the Schuylkill Navigation Company, Lehigh Coal and Navigation Company, and Delaware and Hudson Canal Company, 1820–40.)

could make such a moove in that coal & Navig^n of the Lehigh & thus get out of my difficulties." The difficult terrain in the Lehigh Valley, coupled with the skepticism that surrounded the anthracite trade, however, convinced many Pennsylvania legislators that they had bestowed on the Lehigh Navigation Company a charter for, in the words of one anonymous legislator, "the privilege of ruining themselves." In 1820 the Navigation Company merged with the Lehigh Coal Company, which had been formed to mine the coal and bring it from the hills to Mauch Chunk via a wagon road. The resulting Lehigh Coal and Navigation Company (LCNC) represented a large corporate interest capitalized at $200,000. Endowed with both mining and transportation privileges, it eventually dominated the Lehigh anthracite region. In 1820 the company's first shipments of anthracite coal down the Lehigh amounted to a modest 365 tons, but regular navigation had arrived. Over the next five years the LCNC was the sole improved navigation for anthracite coal, although the SNC caught up quickly once its directors solved their capital and engineering problems.[41]

The competition between the miners using the SNC and the LCNC took on both an ideological and a practical dimension. The ideological component related to their corporate powers and will be treated in more detail later, but the practical aspect of competition among anthracite carriers developed as a result of different mining systems in the Schuylkill and Lehigh regions. The fact that the LCNC held the right to mine as well as to carry coal proved critical, for it allowed the firm to integrate the mining and shipping of coal from the Lehigh area. In 1827 the LCNC established a railroad from its mines at Summit Hill to its terminus at Mauch Chunk—the first railroad to transport coal in the United States. At the same time, the directors of the LCNC also decided to transform their improved river navigation into a slack water canal system. This regulated the level of water and enabled them to allow larger boats to pass through their works. The improvements, coupled with the LCNC's ownership of 8,000 acres of prime coal-bearing land, suggested that the firm could help establish large corporate coal mining in Pennsylvania's anthracite fields during the 1820s.[42]

Despite fears among Schuylkill area miners that the LCNC would manipulate coal prices by alternately starving and flooding the market, actual figures suggest that competition forced both firms to maximize traffic (see fig. 2.1). Since the SNC depended upon toll revenues, not sales, for its revenue, it needed to attract a high volume of traffic in order to earn profits. Its directors kept tolls relatively low in order to encourage coal traffic along the SNC. With more and more Schuylkill region coal arriving in Philadelphia during the late 1820s and early 1830s, the LCNC needed to increase production or risk losing the initial advantage of Lehigh coal among Philadelphia consumers. Rather than tightly control production, the LCNC and other mining firms in the Lehigh region leased out their coal lands to individual miners. Both lessees and the LCNC profited from high production levels, and a system in which the LCNC regulated production in order to maintain high prices never materialized. The result was a steady increase in the supply of anthracite coal reaching Philadelphia from both regions, attended by a steady decrease in price. Unlike the monopolistic position of the James River canal, canals in the anthracite fields competed fiercely for shares of the Philadelphia anthracite market.[43]

The Delaware and Hudson Canal Company (D&H) developed into the third major promoter of anthracite during the early years of the trade. After it received charters from both the New York and Pennsylvania legislatures in 1823, the D&H built a 106-mile canal linking the tidewater on the Hudson to Honesdale, Pennsylvania, in 1828. The D&H then constructed a railroad with one of the nation's first steam engine locomotives to tap into Pennsylvania's northern anthracite field at Carbondale. The firm began shipping anthracite along its route in 1829, and two years later coal completely eclipsed other kinds of traffic along the D&H. Like the LCNC, the D&H had the power to mine, transport, and sell its own coal, but, because New York City represented its major market and the route had little competition during its early years, the D&H encountered less political opposition to its liberal charter than the LCNC did. Supporters of the Delaware and Hudson Canal saw the project as a "developmental line," much in the tradition of New York's Erie Canal. The future of anthracite coal figured prominently in this process, as investors considered New York City's fuel needs with an eager eye. "The introduction of the article among a people to whom its properties are but partially known," one stockholder wrote in 1831, "and whose prejudices, from long habit, may be in favor of wood, will require some time and attention." Following the success of the campaign to promote anthracite, the D&H eventually pried a wedge for anthracite into one of the most important markets for domestic and industrial fuel in the United States.[44]

As the capacity of the anthracite canals increased, colliers stepped up production from the most accessible seams of the northern, middle, and southern anthracite fields. The Schuylkill region's coal mines resembled those of the Richmond basin in structure. Miners usually raised coal under a short-term lease from the land's owner, which encouraged short-term practices that focused upon winning as much coal as possible at minimum expense. The Lehigh Coal and Navigation Company hired its own laborers to raise coal, but it also sub-contracted a portion of its holdings out to small-scale miners. Although deep-shaft mining eventually appeared in both regions, coal mining in the Schuylkill and Lehigh regions during the 1820s focused largely upon seams that lay close to the surface. Much like Virginia's early mines, these enterprises mainly used trenches or quarries to get at the coal. Unlike miners in the Richmond basin, who used slave labor, Pennsylvania colliers often mined the coal themselves or hired white laborers. Some of these hands were recent immigrants from Wales, Cornwall, or England, with practical experience in the art of mining, while others were simply unskilled laborers attracted by the boom mentality of the anthracite regions. Whatever method anthracite miners used to raise their coal, they produced more of it year after year. As long as miners worked seams that lay close to the surface, expanding production was possible by simply hiring more hands to raise the coal. Unlike their cohorts in the Richmond basin, anthracite miners found their coal easier to raise because extensive anthracite mining had only begun in earnest during the 1820s, and many seams remained easily accessible.[45]

By the early 1830s lower prices and reliable shipments allowed anthracite coal to emerge as an important commodity and certainly aided its growth in urban markets. Philadelphia became the first major city to burn anthracite extensively and emerged as the center of the seaboard anthracite trade. In 1825 the managers of the Pennsylvania hospital reported that their purchase of one hundred tons of Lehigh coal had saved them more than a thousand dollars that year. "I cannot recommend it too warmly for general use," wrote hospital steward Samuel Mason, "both in point of economy, comfort, and safety." The increased supply of anthracite and declining stocks of wood during the late 1820s and early 1830s resulted in a general decline of retail prices for coal. Once knowledge of anthracite's special needs circulated among the urban population, anthracite's price became a significant factor in fuel consumption. In 1831, for example, the estimated fuel cost for an inexpensive anthracite coal stove was about $4.50 during the colder months, while the cost of burning wood to heat the same home was about $21. As it became cheap and efficient, anthracite represented the perfect fuel for all levels of society, a fact not lost on Philadelphia's Fuel Savings Society. "The article of Anthracite Coal ought to be introduced as a common Fuel among the poorer classes of our citizens," the society maintained, and it advertised for the development of a cheap apparatus for burning anthracite, "not only for the purpose of warming the apartment, but for doing the necessary cooking, &c. for the family." In the winter of 1831–32 Philadelphia's growing dependence upon anthracite for domestic consumption caused a shortage of stone coal in the city. Many Philadelphians blamed the popularity of cooking with anthracite, which left less stone coal for fireplaces and increased prices. By 1833 Philadelphians consumed over four hundred thousand dollars worth of anthracite, and in 1835 the Pottsville *Miners' Journal* confidently predicted "that coal will in a few years almost entirely usurp the place of wood as a fuel."[46]

Anthracite coal did become a significant alternative to wood or bituminous coal in cities of the eastern seaboard. Demand for this new fuel increased as news of its prodigious heating qualities spread and as more stone coal arrived in cities. In Boston coal retailers sold Pennsylvania anthracite alongside Newcastle and Richmond bituminous, and in 1834 over seventy-six thousand tons of anthracite were shipped there. In Baltimore *Niles' Register* noted in 1831 that anthracite was passing into general use and suggested that "perhaps the cost of a coal fire is equal to one of wood, for the latter is very cheap with us; but the former saves much time, and is the least dangerous." In 1832 the same journal noted that the price of anthracite exceeded that of wood in Baltimore, but stone coal was considered "more convenient, in some cases, and so preferred."[47]

Perhaps the crowning achievement in the campaign to promote anthracite was its successful introduction in New York City. In 1830 New Yorkers burned 23,605 tons of anthracite, which represented nearly a quarter of the total amount spent on fuel in the city. Wood accounted for 60 percent that year, Richmond coal for less than 10 percent, and charcoal a little more than 5 percent. By 1832 over fifty thousand tons of anthracite coal were sold in New York City at an average retail price of $10.65 per ton, which accounted for 38 percent of total fuel sales in New York. When compared to wood (45%), charcoal (7%), and Virginia bituminous (7%), New Yorkers were spending more and more money on anthracite to heat their homes. The increased dependence upon anthracite is evident also in the fact that New Yorkers suffered a shortage of anthracite during the winter of 1831–32. Despite the suffering that occurred as a result of the shortage, sales of anthracite coal in New York City continued to grow, and in 1832 the *New York Constellation* expressed affection for burning anthracite in verse:

> a reddish brown, soft, fine impalpable;
> And if the fire once lit, continue long,
> Glowing and lively, sending forth the heat;
> The Coal is good and fit to warm the hearths,
> Of honest men. Make haste to purchase more,
> If more there be, and you are not supplied.[48]

The most impressive aspect of this upsurge in anthracite consumption in New York City is that it began at a time when stone coal was still more expensive than Virginia bituminous. Table 2.2 demonstrates that the wholesale price of anthracite at the height of the heating season remained almost double that of Richmond bituminous. Solid comparative data on wholesale prices are not available for the years after 1831, but best estimates suggest that the price per ton of anthracite at Philadelphia still averaged a little under six dollars a ton from 1834 to 1840. Prices dropped to under four dollars per ton during the 1840s, but by that time few Philadelphians or New Yorkers still burned Richmond basin coal. A drop in prices made anthracite coal affordable to urban consumers, although price levels do not fully explain anthracite's rise.[49]

By the 1830s anthracite coal's domestic utility was well known, and de-

Table 2.2. Wholesale Price per Ton of Coal in New York City, 1827–1831, in Dollars

	1827	1828	1829	1830	1831
Richmond bituminous	6.71	6.34	5.92	5.92	5.09
British bituminous	10.79	12.86	14.28	11.20	8.83
Pennsylvania anthracite	12.00	11.50	11.50	12.00	7.50

Source: Cole, Wholesale Commodity Prices in the United States, 1700–1861, vol. 2: Statistical Supplement, 224–40.
 Note: January quotes are used here to reflect peak demand for both domestic and industrial fuel.

mand for the product increased annually. The resulting boom in Pennsylvania's anthracite regions, especially in the Schuylkill region, ensured that stone coal would never again be in short supply, as it had been in the winter of 1831–32. Throughout the decade a meteoric rise in the price of coal lands and anthracite canal stocks astounded observers. In 1835 *Niles' Register* commented on the rising values of land: "As to the lands—persons do not know what to ask for them—they suppose the whole union will become tributary to them!" A year later the *North American Review* reported that, "a new world seemed to have sprung up in the wilderness, as if by enchantment." The mountains of eastern Pennsylvania represented the center of a new industry that "converted the wildest waste into the theatre of active life, given a fresh stimulus to individual enterprise, created an inexhaustible source of wealth in which it lay, and opened a new commerce and a new bond of fraternity to the whole Union." The value of the three main anthracite canals also soared. By 1835 the Schuylkill Navigation Company's stock was selling for $145, nearly three times its $50 par value in Philadelphia, while the Lehigh Coal and Navigation Company's $50 shares were selling for $90.50 and the Delaware and Hudson's $100 shares for $121.[50]

Although pleased with the rapid growth of their trade, boosters of anthracite envisioned even more lucrative markets in the future. New uses for anthracite coal, particularly in the United States' growing iron industry, would increase the demand for stone coal and enrich the miners of the Schuylkill, Lehigh, and Wyoming regions. But, more important, the increased utility of anthracite became a statewide concern. Widespread support for the anthracite coal industry emerged as a patriotic cause for Pennsylvania's state government because it appeared exclusively within the commonwealth's boundaries. In 1826 Governor John Andrew Schulze articulated this link between commonwealth and coal when he called for prompt legislative action to facilitate a "rapid development of resources, of which the importance and extent were till lately, neither appreciated nor understood." "Our mountains and waste lands, which seemed to be doomed to everlasting barrenness," Schulze argued, "are found to teem with inexhaustible wealth, and to constitute, if not the fairest, certainly the richest portion of our territory."[51]

The state first threw its weight behind the anthracite campaign when the question of taxing this new mineral arose in response to a fiscal crisis in the early 1830s. Pennsylvania's primary economic program, its ambitious network of state-owned and -operated canals, pushed the state to the brink of financial ruin. Since 1823 the commonwealth had paid for its internal improvements through borrowing rather than levying taxes upon its population. As a result, the state's funded debt had increased to an estimated $12.5 million by 1831, with annual interest payments at $616,850. After paying for the normal costs of government, the Pennsylvania treasury could only pay for about two-thirds of the interest on its debt that year. Taxation, which for so long had been a political impossibility, now became a necessity if the commonwealth were to remain solvent. The administration of Governor George Wolf foresaw this fiscal crisis and floated a number of tax proposals to the legislature in January 1830. Among them were property taxes aimed largely at the rich, levies upon inns and taverns, an increase in county taxes, and a light tax on anthracite and bituminous coal of twenty-five cents a ton.[52]

Needless to say, the proposed taxes created quite a political stir all over Pennsylvania and especially in the anthracite regions. Agitation against the coal tax was to a degree inspired by the belief that some citizens would end up paying for state internal improvements that benefited others, but most specific opposition to the coal tax was based on the idea that it would stunt the commonwealth's economic growth. In March 1830 a large meeting of Schuylkill County residents in Pottsville passed a series of resolutions against the tax peppered with patriotic rhetoric. "The tax will operate as a bounty upon the Coal of other states," declared the petitioners, and will "enable the coal traders of Nova Scotia, Rhode Island, Maryland and Virginia, who are free from this

impost, to undersell our Coal in a foreign market; and thus confine our product to the Pennsylvania market alone." Therefore, they argued, "the tax would
be fatal to the competition of our coal trade and of our manufacturers with
those of other states." By the time the legislature convened in the winter of
1830, an anti–coal tax movement was well in place. Petitions flooded into the
legislature in January and February 1831. Schuylkill County led the way with
twenty-five separate remonstrances, but other counties in eastern and central
Pennsylvania also sent petitions against the tax.[53]

The Committee of Ways and Means invoked no small measure of Pennsylvania patriotism in recommending against the coal tax in January 1831. The
tax might be a good way to raise revenue, the committee concluded, but was
bad for all Pennsylvanians and, indeed, all Americans. The committee argued
that in its very principle the tax was "unequal and oppressive in its operation"
and operated as both "a tax on the poor" and "a bounty to the labor and enterprize of citizens of other states, where coal lands are situated." A program
of internal improvements required increased revenues in order for the state to
remain solvent, the committee concluded, but such a valuable asset as the coal
trade should not be sacrificed in the name of fiscal responsibility. The committee also recognized the value of anthracite in out-of-state markets such as
New York, a point that several petitions raised in the campaign against the coal
tax. In response, the House of Representatives defeated the coal tax bill in February 1831. A breakdown of the vote reveals that the apportionment of Pennsylvania's legislature by population made a major difference in defeating the
coal tax, as delegates representing the Philadelphia area constituted nearly a
quarter of the fifty-four nays, while the thirty-eight yeas were centered mostly
in rural counties that did not contain elements of the anthracite or bituminous coal mining industry. Philadelphia interests provided the swing vote, for,
had they supported the tax, the issue would have passed. In general political
opposition to the coal tax came from all regions of the state, but urban delegates provided critical votes.[54]

The importance of this issue lies not simply in the tax's defeat but also in
the arguments used to organize opposition to it. Opponents of the tax de-
ployed a strong, and ultimately convincing, rhetoric that explicitly linked the
future of Pennsylvania with the future of anthracite. In the 1830s anthracite

coal became ingrained in the politics of the Keystone State. Governor Wolf ultimately withdrew his support for the coal tax in the face of vocal opposition, but the very idea of a shortsighted executive who would contemplate retarding the growth of the coal industry with a tax created major problems for Wolf's campaign in 1832. Gubernatorial candidate Joseph Ritner argued as much. "The true policy in regard to this valuable mineral," he wrote in a letter circulated during his campaign against Wolf, "would be, instead of throwing legal obstructions in the way of its being brought into general use, to afford every facility to bring it to market." Indeed, following the defeat of the coal tax in 1831, the idea that the state should make every effort to promote the use of anthracite encountered little real opposition. "There is no reason why Pennsylvania should not attach to her Coal Trade the same importance as Great Britain does," argued one correspondent to the *Miners' Journal*, and, as the decade progressed, many Pennsylvanians found the economic prospects of their state more and more linked to stone coal.[55]

Comparison with the British and Pennsylvania coal trade, moreover, led many to think that Pennsylvania should emulate the British in using mineral fuel to make iron. Since the 1780s bituminous coal or coke had been the preferred fuel for British iron makers. Once anthracite had begun to enter American hearths, there seemed to be no reason why stone coal could not be used to make iron. As with its domestic use, however, the industrial potential of anthracite coal faced major technological barriers. In British and American iron furnaces of the early nineteenth century, the high heat needed to smelt iron ore required a blast of excess air to aid the combustion of the fuel, whether it was coal, wood, or charcoal. While British iron makers in the 1820s attempted to increase the efficiency of the process by using superheated air, known commonly as a "hot blast," American iron makers still used a "cold blast" to stoke their furnaces. Because anthracite is so dense, however, it resisted attempts to ignite it with the cold blast method. Long-established patterns of charging an iron furnace could not, for example, accommodate the particular combustive conditions required by anthracite. Although blacksmiths in anthracite mining districts adapted their small-scale furnaces to this type of coal by installing grates and separating the fuel from its ashes, conventional iron furnaces in rural Pennsylvania could not be so modified. Stone coal at first glance appeared to be an inappropriate fuel for most American iron furnaces.[56]

Some financial incentives to develop a method of burning anthracite coal in iron furnaces arose from the private sector during the 1820s and 1830. For example, the Franklin Institute joined the ranks of anthracite iron enthusiasts with its established program of technical articles, descriptions of experiments, and premiums at its nationally renowned industrial exhibitions. During the 1830s the Lehigh Coal and Navigation Company promised free waterpower and cheap coal to a company that could consistently produce iron with anthracite, and Philadelphia's Nicholas Biddle offered a cash prize of five thousand dollars for the same function.[57]

In order to make anthracite iron a reality, however, Pennsylvanians enlisted the aid of their state government. In doing so, they tapped the great potential of state-level institutions to underwrite private industry, as the prospective wealth to be gained from iron smelted with anthracite inspired Pennsylvania's state officials. But the state needed to do more than simply wait for anthracite iron to develop in the private sector, as iron making in the nineteenth century was a capital-intensive industry. Convincing a family to buy an anthracite heating grate for its fireplace or a cooking stove might be within the capabilities of private institutions; sponsoring an entirely new iron industry was not. "The legislature may do essential good *in saving time,*" the *Miners' Journal* argued in 1831, "by allowing ten or twenty thousand dollars premium to him or them who will produce the first five hundred tons of iron equal to English, and manufactured of mineral coal, or they may encourage individual enterprize and capitalists at the commencement, by acts of incorporation, with liberal privileges." Either way, the state would simply have to be involved with the promotion of anthracite iron.[58]

Direct financial assistance in the form of bounties and premiums seemed like a good idea, but during the 1830s it was impractical for Pennsylvania's legislature. The state was already fiscally strapped, so the direct financial backing of anthracite iron was impossible. Here the open structure of Pennsylvania's legislature worked against the state support of the coal trade, as competing interests killed the idea of awarding cash bonuses for anthracite iron. In 1838

Representative J. H. Laverty of Clearfield County introduced a motion to award a premium to two entrepreneurs from his district for "their valuable efforts and success in the manufacture of iron from the ore with coke or mineral coal." Most legislators could agree, as the resolution asserted that "it is right that such enterprize should be encouraged and rewarded by some signal act of public approbation by the government." But they could not agree on the recipients of this patronage. Laverty's bill was stalled when an amendment extended a premium to F. H. Oliphant of Fayette County as well. The next year two state senators from Philadelphia introduced a bill to establish an award for every hundred tons of iron manufactured exclusively with anthracite or bituminous coal, but they failed to gain enough legislative support.[59]

In the end the Commonwealth of Pennsylvania took an active role in promoting anthracite-smelted iron without directly funding technological innovation. In 1836 the legislature passed an act entitled "To Encourage the Manufacture of Iron with Coke or Mineral Coal, and For Other Purposes," which allowed any iron-making firm that met certain requirements to receive a corporate charter so long as it burned coal, rather than charcoal, as fuel. The act required incorporating firms to have at least $100,000 in capital and no more than $500,000. It also allowed them to hold up to two thousand acres either in a single county or in two adjacent counties. Enacting a general incorporation law was a potentially controversial way to sponsor anthracite iron, since the right to incorporate manufacturing enterprises in Pennsylvania was jealously guarded by the legislature during the 1830s. The usual practice of creating corporations (which will be discussed in more detail in chap. 5) involved the passage of a charter by the state legislature—a political process that many public officials found distasteful and inherently corrupt. In fact, Governor Ritner announced in 1836 that corporations often reflect a "depraved appetite" for speculation, "foster and perpetuate the thirst for gain without labor," and paralyze individual entrepreneurs. The bill received minor opposition in the legislature from anti-charter forces and established charcoal iron interests. Supporters cast its intention as a bounty on the fusion of coal and iron, thereby avoiding accusations that the act would create monopolistic corporations. The bill's extension of charters to bituminous furnaces undoubtedly convinced representatives from western and central Pennsylvania to support the bill, even though it mainly addressed anthracite interests.[60]

Anthracite iron became a reality in 1840, when David Thomas brought Welsh hot blast technology into practice at the Lehigh Crane Iron Company, a firm created in 1839 under the general incorporation act. The Lehigh Crane Company's innovation near Allentown created a stir in iron-making circles, and iron furnaces for smelting ore with anthracite began to appear across eastern and central Pennsylvania.[61] In 1841, only a year after the Lehigh Company's success, the chemist and geologist Walter Johnson found no less than eleven anthracite iron furnaces in operation in Pennsylvania—at least three had been created under the 1836 act. He concluded that anthracite iron furnaces "await but the hand of patient industry to render them available, for establishing our ancient commonwealth on the very pinnacle of prosperity."[62] That same year an American correspondent of London bankers cited savings on iron making of up to 25 percent after the conversion to anthracite and noted that "wherever the coal can be procured the proprietors are changing to the new plan; and it is generally believed that the quality of the iron is much improved where the entire process is affected with anthracite coal."[63]

As a result of technological innovation, entrepreneurial effort, and state promotion, anthracite iron eventually dominated U.S. iron production. By 1844 anthracite iron was the cheapest iron made in the United States, and the capacity of Pennsylvania's anthracite furnaces was critical in the dramatic rise of American iron productivity throughout the entire decade of the 1840s. Just as stone coal dominated domestic fuel markets, anthracite furnaces rapidly assumed leadership in American pig iron production. From 1830 to 1860 the railroad network of the United States grew tenfold, which created an unprecedented market for iron products such as rails, cars, and other railroad components. U.S. firms scrambled to satisfy the ravenous appetite of railroads for iron but at first lost ground to cheaper British imports, especially in the critical iron rail market. Tariff levels remained relatively low during most of anthracite iron's rise, which suggested that technology, not necessarily protective duties, helped many American iron firms compete with rail imports.[64]

Pennsylvania's investment in anthracite iron paid dividends for the industrial economy of the state. By 1854, 46 percent of all American pig iron had been smelted with anthracite coal as a fuel, and six years later anthracite's share of pig iron was more than 56 percent. Perhaps the best indication that anthracite iron owed its success to political as well as technological factors is its relatively short period of supremacy. Beginning in the 1870s, the preferences of iron makers swung away from anthracite in favor of bituminous coal and coke, as Pennsylvania's rich Connellsville coking fields came into use. By 1896 anthracite iron accounted for only 1 percent of the nation's pig iron production.[65]

Anthracite iron's success story during the early nineteenth century cannot be solely attributed to Pennsylvania's general incorporation act of 1836 or to the rejection of a tax. By 1846 only three firms that had organized under the 1836 general incorporation act remained among the state's thirty-six anthracite furnaces and accounted for about one-fifth of the total capacity of the state's anthracite iron industry.[66] Entrepreneurs anxious to profit from anthracite iron never limited themselves to general charters and in fact utilized a variety of organizational structures, including individual proprietorships, limited partnerships, and corporations chartered by the legislature. Yet the state's early sponsorship of anthracite through policies such as the 1836 general incorporation act created an institutional environment favorable for the growth of anthracite as an industrial fuel. Moreover, the debate on the proposed coal tax of 1831 suggested that coal had assumed a special place in Pennsylvania's political economy, thus providing the capstone of state legislative support to a promotional campaign that united entrepreneurs, scientists, and politicians in the support of anthracite coal.[67]

A PATTERN ESTABLISHED

What did the Richmond basin's coal industry do while anthracite won over the hearths and furnaces of the eastern seaboard? In short it did not respond at all. Virginia had no counterpart to Philadelphia's Franklin Institute, and its scientific community showed little or no interest in coal. State legislators may have demonstrated empathy for Virginia's miners in the abstract, but they offered little in the way of practical support. Harry Heth's representative in the Virginia General Assembly, for example, asked that Heth not "scrutonize too nicely, and severely" when he voted against Chesterfield County coal interests. At the same time that Pennsylvania anthracite emerged triumphant, activity in the Richmond basin could only be termed sluggish. In 1841 *Hunt's Merchant Magazine* reported that the Virginia coal trade "has been carried on in an easy, careless, unenterprising manner, so characteristic of that venerable common-wealth, that we can scarcely wonder that it has grown into the currency of a proverb, that 'Old Virginny never tires.' The sensation is rather produced in the observer of her movements."[68]

The Richmond basin's position in the U.S. coal trade suffered a relative de-cline from the late 1820s onward. Pennsylvania anthracite's reputation rose, and its share of urban markets increased, while seaboard consumption of Richmond basin bituminous coal shrank, as demonstrated in figure 2.2. After 1836 urban consumers burned less Richmond bituminous than British coal. A brief ten years after New York, Boston, and Philadelphia wholesalers sent Richmond area collier Harry Heth a barrage of requests for his coal, the Rich-mond basin's position in the U.S. coal trade dropped precipitously. Without a plan to counter Pennsylvania's promotional campaign for anthracite, Rich-mond colliers found themselves squeezed out of urban markets.

To add insult to injury, anthracite testimonials often accentuated the phys-ical differences of the two types of coal. Most devastating was the common critique that Richmond bituminous was susceptible to spontaneous combus-tion. In 1828 the *Register of Pennsylvania* reported a number of cases of coal igniting itself in storage rooms, and each time it was careful to note that "Vir-ginia coal" was the culprit. Ten years later the *American Journal of Science* re-ported a similar incident with the headline "Another Case of the Spontaneous Combustion of Virginia Coal." Without a rival scientific or technological com-munity to counter, or at least address, these claims, the image of Richmond coal suffered as that of Pennsylvania's anthracite soared.[69]

The case of Pittsburgh, moreover, demonstrated that, despite the experi-ence of the Richmond basin, local deposits of bituminous coal could become a valuable asset to a growing city. When the War of 1812 disrupted trade be-tween western Pennsylvania and coastal cities, Pittsburgh's manufacturing economy expanded by using cheap and plentiful bituminous coal. The value of iron production in Pittsburgh by 1815, for example, stood at seven times its

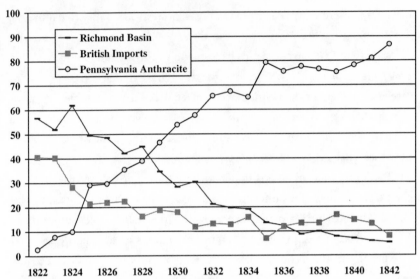

Fig. 2.2. Percentage of seaboard coal consumption by origin, 1822–1842.
(*Hunt's Merchant's Magazine and Commercial Review* 8 [June 1843]: 548; Alfred
Chandler, "Anthracite Coal and the Beginnings of the Industrial Revolution," 154.)

1810 levels. Glass and brass production witnessed similar expansion—based
mainly upon the use of local coal. Annual bituminous coal production in
western Pennsylvania eclipsed 200,000 tons by 1826, and ten years later miners there raised over 450,000 tons of coal. Although Virginia coal suffered from
a reputation of low quality, the experience of Pittsburgh suggests that geological differences between bituminous and anthracite coal were not the culprit
in the Richmond basin's decline.[70]

Without the institutional backing that characterized the rise of anthracite,
Virginia coal miners such as Harry Heth could not compete with their Pennsylvania counterparts in eastern markets. The lack of a technological community and of active state support plagued the Richmond basin throughout
the nineteenth century but proved particularly devastating in the early
decades of the century, given the initial advantage enjoyed by producers of
Virginia bituminous. Annual production in the Richmond area surpassed
200,000 tons once, in 1835, but declined during the 1840s and 1850s to hover
between 100,000 and 140,000 tons annually. It was not until the years follow-

ing the Civil War that mining engineers "rediscovered" Richmond coal and proclaimed it an excellent fuel. The coal's bad reputation, they reasoned, owed more to inexperienced mining methods utilized by local firms than to its geological character. This realization arrived too late, however, for the Richmond basin trade; it never regained the advantageous position it had held at the beginning of the nineteenth century.[71]

The first few decades of competition between anthracite and Richmond bituminous in the U.S. coal trade established that private and public actors could alter the demand for different ranks of coal. As the nineteenth century progressed and the demand for coal increased in both size and scope, colliers in both Virginia and Pennsylvania repeatedly sought the aid of their respective state governments. The rise of anthracite and decline of the Richmond basin during the 1820s and 1830s set a pattern of dynamic growth in Pennsylvania and stagnation in eastern Virginia. Colliers in other regions of Virginia sought to ensure that their fortunes would take a different path from their colleagues in the Richmond basin. Perhaps, they hoped, the aid of state government could break the pattern of lethargy, just as Jacob Cist, the Franklin Institute, and public incentives had broken the Richmond basin's advantage in the early U.S. coal trade. As the story of internal improvements, geological surveys, and corporate charters in both states demonstrate, political conditions in Pennsylvania and Virginia reinforced, rather than redirected, the paths that the states' coal trade took in the years following the War of 1812.

Trunk AND Branch

State Internal Improvement Networks and the Coal Trade

I n the summer of 1828 an internal improvements convention in Charlottesville attracted over 125 influential Virginians, mostly drawn from counties along the James River. With former president James Madison as chair of the convention, the delegates debated the future course of state-aided transportation in Virginia for six days. Many heaped scorn on the Old Dominion's existing system of supporting a number of small-scale, but politically prudent, turnpike and canal companies. The consolidation of legislative support for projects, therefore, emerged as a central theme of the convention. Although the delegates in Charlottesville recommended that the legislature continue to support existing projects such as the improvement of the Shenandoah, Roanoke, and Potomac rivers, the most emphatic endorsement was for the completion of the entire James and Kanawha system. Not only did this planned improvement "pass in its whole extent through the centre of the State, and accommodates a greater number of our fellow citizens than any other river," argued the delegates, it also "promises more than any other, to be auxiliary to the extension of the commerce of Virginia and the West." With the recent reorganization of the James River Company, moreover, the state controlled the length of the improvement. Delegates predicted that the tolls from this centerpiece project would enrich Virginia's treasury and support other works. "Virginia entered upon this course with eagerness," they argued, "[and] pursued it for some time with animation." "Will true wisdom justify the continuance?"[1]

Delegates in Charlottesville drew upon a remarkable blueprint for state-funded internal improvement networks. In 1825 New York's Erie Canal linked the Hudson River at Albany to Buffalo and the Great Lakes, only eight years after work on the project had begun. The fruits of the endeavor proved both impressive and immediate. In the first year of its operation toll revenues on the Erie Canal surpassed the annual interest on the state's construction debt, as traffic on the improvement ranged from heavy freight such as lumber and wheat, small manufactured valuables, and passengers using the canal for both speedy transportation and leisure. By 1837 the revenues from the Erie Canal erased New York's construction debt completely. The waterway also shortened the time and expense required for the transportation of both bulk and high-value commodities considerably. Freight rates from Buffalo to New York City fell from an average of one hundred dollars a ton to less than nine dollars a ton during the 1830s. This opened up New York's western counties to development, and the growing cities of Buffalo, Syracuse, and Rochester all prospered from bordering the Erie Canal. Moreover, as a public works project constructed by New York's state government, the Erie Canal demonstrated the potential of a state-funded internal improvement network to benefit economic development.[2]

Both Virginia and Pennsylvania rushed to copy the success of New York's canal experiment, and the interaction of interests in the formation of internal improvement policy reflected the divergent paths of the policy-making systems of these two states. Coal also figured prominently in the construction of both systems. In Pennsylvania the open structure of the legislature and the need to compromise allowed a number of competing interests to influence the character of the State Works. Unexploited anthracite and bituminous fields figured in this political calculus, and the resulting public improvements aided their development by providing their initial linkages to market centers. Political pressure from existing anthracite concerns also kept most of Pennsylvania's public improvements from competing directly with the private coal carriers. In dealing with a myriad of groups interested in the construction of public works, Pennsylvania's responsive state government allowed development to occur without major setbacks.

In contrast, policy making in Virginia's legislature forced all minority interests, including the coal interests, either to adapt to the James River and Kanawha system or to attempt to construct transportation links privately. The James River system thus hindered development of the rich bituminous regions of both the tidewater and the trans-Allegheny region, while Virginia's paucity of investment capital ruled out private canals. Unlike its counterpart in Harrisburg, the Virginia legislature could not accommodate competing interests in regards to internal improvements. Thus, it championed a narrow vision of the James River and Kanawha route. This stark contrast in internal improvement policy furthered the gap between Pennsylvania anthracite and Richmond bituminous and retarded the development of the Old Dominion's western bituminous region, while Pennsylvania's bituminous coal trade flourished.[3]

EAST VERSUS WEST: THE JAMES RIVER AND KANAWHA CANAL AND SECTIONAL POLITICS

Much like the Richmond basin, Virginia's western coalfields required extensive investments in transportation in order to increase production and reach urban markets. In the Kanawha Valley local salt producers burned a significant amount of coal in their furnaces, but the Ohio Valley trade promised even greater rewards for local colliers. In 1818, for example, a Cincinnati merchant estimated the annual use of coal between the mouth of the Kanawha and the Falls of the Ohio at 116,000 bushels. The successful trip of the steamer *Andrew Donnally* up the Kanawha River to Charleston in 1820 heralded the arrival of steamships as both a consumer and carrier of mineral fuel. Coal mining operations in the Kanawha region began as small-scale efforts designed to sup-

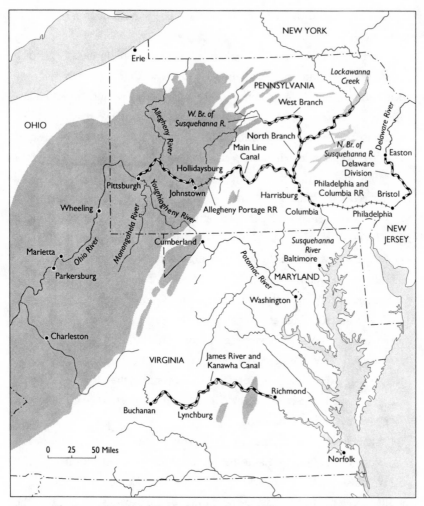

Map 4. The James River and Kanawha Canal and the Pennsylvania State Works

plement dwindling wood stock for the salt furnaces of the area but eventually grew to become the salt industry's main source of fuel. Colliers in western Virginia usually employed less than ten miners and used African-American slaves in their operations as both skilled miners and general laborers. Because the seams first exploited laid close to the surface, few mines required drainage pumps or ventilation systems in the early years of the Kanawha coal trade. Many of these operations were "captive" mines, meaning that salt manufacturers either owned them outright or leased coal lands out to independent contractors.[4]

In many ways the Kanawha coal trade followed the pattern established by the neighboring bituminous coal industry to the north in the Pittsburgh area. Local entrepreneurs began mining coal to satisfy local demand; they then discovered that the region contained a number of profitable seams and aspired to increase production in order to export coal from the region. Only one critical difference between the Kanawha Valley and Pittsburgh endured. At the same time that coal mining became an essential ingredient of the Kanawha Valley's economic future, its lifeline to the Ohio-Mississippi system, the Kanawha River, remained basically unimproved, and the region was isolated from the rest of Virginia. Regular navigation, especially for coal boats weighted down with their bulky cargo, needed deeper drafts than the forces of nature on the Kanawha usually allowed. In addition to exploiting western markets, colliers of the Kanawha Valley also envisioned opportunities in the growing manufacturing centers of central and eastern Virginia. If somehow the rich mineral resources of western Virginia could be linked to the valley and the tidewater, they reasoned, the entire state could benefit from the trans-Allegheny's natural wealth. A statewide improvement, therefore, would provide an indispensable boost to the Old Dominion's western coal trade both within and beyond its boundaries.

Colliers in western Virginia quickly discovered that their hopes for developing coal traffic along the Kanawha resided in the same institution that their eastern colleagues found so lacking: the James River Company (JRC). But, as the JRC evolved into a statewide endeavor, the firm became enmeshed in the affairs of Virginia's conservative polity. In 1812 the legislature appointed a commission to explore the possibility of connecting the James River to the falls of the Kanawha via the Greenbrier and New rivers. The commission concluded that Virginia's "central channel of communication of water may be of great value" to the economic future of the state. Plans to build a system based upon this report percolated in the legislature, but the outbreak of the War of 1812 delayed any potential public action. Four years later the General Assembly set the ball rolling again when it passed an act creating an Internal Improvement Fund for "rendering navigable, and uniting by canals, the principal rivers, and of more intimately connecting, by public highways, the different parts of this Commonwealth."[5]

As visions of a tidewater to Ohio River system gained steam in legislative circles, the James River Company limped along in its promise to assume statewide dimensions. By 1820 frustration with the firm's inability to develop its navigation along the length of the James forced a reorganization of the firm which placed it under state control. Under the new arrangement the James River Company would be responsible for the improvement of the Kanawha River from the Great Falls to the mouth of the Ohio, the construction of a road from the Great Falls to the mouth of Dunlop's Creek in western Virginia, the creation of a navigable canal from Dunlop's Creek along the James River to Pleasant's Island, and the improvement of the navigation of the James from Pleasant's Island to tidewater. The legislature also mandated that the James River Company would become an agent in trust, holding the works for the benefit of the commonwealth, and that state officials would annually appoint nine commissioners to oversee both the eastern and western ends of the works. A private board of directors, however, would maintain operational control over the firm. In return for relinquishing much of its supervisory authority, Virginia guaranteed the original stockholders a dividend of 12 percent on the par value of the stock for twelve years and 15 percent afterward. Three years later the legislature replaced the existing corporate president with the governor and the private board of directors with Virginia's lieutenant governor, treasurer, auditor of the commonwealth, and the second auditor. The General Assembly also appointed separate commissioners for the James and the Kanawha improvements. Most important, all loans incurred by the firm were now backed by the full faith of the Commonwealth of Virginia, and the Board of Public Works still held 427 of the James River Company's 700 shares. For all intents and purposes the James River Company was a completely public concern by 1823.[6]

Despite this turn of events, industrial development in western Virginia still faced significant opposition. At the same time that political support for internal improvements shifted from local projects to statewide endeavors, the white population among Virginia's four major regions was evenly distributed. But, in terms of representation, the tidewater sent seventy-five delegates to the

General Assembly along with seven senators. The piedmont sent fifty-eight delegates plus eight senators to Richmond. All in all, eastern Virginia held 62 percent of the legislative seats. The valley sent twenty-eight delegates and four senators to Richmond, and the trans-Allegheny west claimed fifty-two delegates and five senators, which constituted 38 percent of the legislature. Nearly two-thirds of the House of Delegates and Senate thus came from the slave-holding east. By 1828 a majority of enfranchised white Virginians favored a constitutional convention to redress these changes. Western interests thus remained skeptical of their clout in the state's internal improvement schemes, and the shape of the future General Assembly was very much in doubt.[7]

The constitutional convention of 1829–30 quickly split along sectional lines. Eastern delegates favored representation on a mixed basis of the free and slave population based upon the federal constitution, but westerners demanded that the General Assembly only reflect the number of white male voters in the state. Furthermore, Virginia's limitation of suffrage to white male freeholders, or those owning more than fifty acres, automatically disqualified between two-thirds and three-fourths of the free white male population. The percentage of whites who actually voted in elections hovered between 6 and 9 percent of Virginia's white population before 1851. In light of this fact, western delegates sought to allow universal white male suffrage in the state, which would not only widen the franchise among white Virginians but would also increase the west's voting power.[8]

Throughout the three and a half months of the 1829–30 convention, eastern delegates took a hard-line stance on all issues that even hinted at undermining slavery's future in the Old Dominion. Although the argument hinged largely upon universal white male suffrage as the basis for apportionment, another fear of the eastern delegates was that tax dollars would flow from the east to expensive internal improvement projects in the west. The convention became a kind of last hurrah for a number of prominent Virginians such as James Madison, James Monroe, and John Marshall. In the end, however, it was the staunch conservatives, led by John Randolph of Roanoke and Benjamin Watkins Leigh of Richmond, who dominated the convention and ensured that it achieved very little in the way of reform. These eastern conservatives pushed for a mixed basis of representation, in which the legislature would be appor-

tioned according to a formula that combined the white population of the state
with three-fifths of the black population to determine representation. Reform
delegates, on the other hand, sought representation based solely upon Virginia's white population. After considering a number of compromise plans,
the convention eventually adopted the recommendations of a select committee to apportion the legislature using an average of the white and the mixed
basis.[9]

The committee made a slight adjustment to the regional breakdown of the
General Assembly, but the balance of power still rested east of the Blue Ridge.
The select committee reduced the House of Delegates from 212 to 134 members and increased the Senate from 24 to 32 members. It also ruled that reapportionment would require a vote of two-thirds of both houses of the General Assembly and could not be attempted before 1841. In April 1830 the new
constitution was ratified by a vote of 26,055 to 15,563, even though it was re-
soundingly defeated in areas west of the Blue Ridge. Western Virginians found
little to cheer about in the compromise, and the conservative framework created by the Constitution of 1830 hardly abated their calls for universal white
suffrage in Virginia over the next two decades. Universal white suffrage eventually was granted through another constitutional convention in 1850–51, but
the General Assembly's apportionment on a mixed basis of slave and free population—a method that empowered eastern slaveholding counties and weakened western interests—continued throughout the antebellum era.[10]

The future of western Virginia's coal trade, therefore, resided in the hands
of a legislature dominated by eastern slaveholding interests. Because no other
transportation project was likely to receive political backing and financial support, residents of the Kanawha Valley depended upon the state-funded James
River Company to provide their outlet to western markets. During the early
1820s salt manufacturers, frustrated with the inactivity of the James River
Company's improvements to the Kanawha River, petitioned the legislature to
allow the Baltimore and Ohio Railroad to build a line through the Kanawha
Valley to the Ohio. Eastern delegates, who successfully restricted the B&O's
western terminus to a point north of the mouth of the Little Kanawha in 1827,
rebuked them. Sectional votes in the legislature also defeated the attempts of
the Staunton and Potomac Company and the Lynchburg and New River Rail-

road to connect the region to the Ohio Valley. Without any viable alternatives, Kanawha region colliers found themselves stuck with the James River Company.[11]

Western colliers had reason to look for alternatives, as the James River Company's actions never lived up to the region's needs. The reorganized James River Company undertook three simultaneous projects to complete navigation from the Ohio to tidewater. The first connected Richmond to Maiden's Adventure Falls, about thirty miles up the James River. The second section was the Kanawha Turnpike, which connected Charleston to Covington, a town in the Blue Ridge Mountains. The third section included the improvement of the Kanawha River to the Ohio. The Board of Public Works expected that revenues from the coal trade would defray the cost of the first section, revenues from the turnpike would pay for the second, and salt tolls would fund the Kanawha River improvement. Although the James River Company eventually completed all three projects, the assumption that revenues would exceed costs proved unrealistic, and expectations that the James-Kanawha waterway would rival the Erie Canal went unfulfilled during the antebellum era.[12]

Of the three divisions of the state-run James River Company, for example, the Kanawha improvement was the least effective in attracting traffic. The first improvements of the Kanawha River consisted of enlarging the channels through the use of wing-dams and sluices to create a deeper navigation between larger basins already existing in the river. By 1830 this navigation was completed from Charleston to the mouth of the Kanawha at Point Pleasant, a distance of some fifty-eight miles. Yet work on the river did not create a draft deep enough to sustain regular steamboat traffic along the Kanawha, and it was definitely inadequate for transporting large amounts of salt or coal to the Ohio. In addition, the company spent more than twice as much on the Kanawha Turnpike as it did on the river navigation—a point that irked many colliers and salt manufacturers along the Kanawha. The experience of Richmond basin miners had already demonstrated that turnpike roads could not sustain a growing coal trade, and residents in the Kanawha Valley complained that their needs were hardly being met. Clearly, the salt and coal trade required more extensive renovation in the way of deeper channels and wider navigations than the state-controlled James River Company was willing to provide.[13]

Throughout the 1820s and 1830s the antipathy of westerners toward internal improvement policy in Richmond persisted. Kanawha Valley salt producers faced a major recession when the price of salt plummeted from thirty-seven cents a bushel in 1820 to twelve cents in 1826. The high costs imposed by turnpikes, combined with the undependable Kanawha River navigation, threatened the future of salt exports from the region. As the salt industry stalled, colliers hoped to continue production for export markets along the Ohio and Mississippi valleys. But, despite their numerous petitions to the legislature and demands for relief, the state-run James River Company continued to favor the eastern half of its improvements. A House of Delegates committee on roads and inland navigation, for example, suggested that tolls on the eastern portion of the James canal be increased so as to pay for the western portion of the canal, but eastern delegates defeated the higher tolls through a sectional vote. In their opinion the James and Kanawha rivers improvement was a "national scheme" cooked up by the Board of Public Works and that "to extract from the commerce of their constituents an increased surplus of tolls, for the purpose of relieving the general funds of the state, was to violate the clearest principles of justice and equality, by visiting exclusively upon *a part*, burthens equally appertaining to *the whole*."[14]

Because the James River Company held the improvement in trust for the state, western Virginians saw the same forces that had squelched reform in the constitutional convention—namely, eastern slaveholding interests—as being responsible for their poor economic position in the Ohio Valley trade in salt and coal. Estimates on funding supported this criticism. The Second Auditor of Virginia concluded that, of the $1.3 million expended by the state on the James and Kanawha rivers through 1831, nearly 80 percent had gone toward the improvement of the James River from tidewater to the Blue Ridge. Although eastern Virginians claimed that because they provided the bulk of tax revenues they should receive the majority of funding, such shortsightedness seemed contrary to the spirit of successful developmental lines such as the Erie Canal in New York. Westerners responded to this perceived disparity with disgust and outrage. "Disregard the claims of the trans-Allegheny counties to what they may deem a proper share of the funds of internal improvement," the *Winchester Republican* warned in December 1830, "and *a division of the state must follow*—not immediately, perhaps, but the signal will be given for the rising of the clans, and *they will rise.*"[15]

Because the James River Company failed to link the Ohio River with Virginia's tidewater, supporters of a statewide canal urged the commonwealth to assume even more organizational and financial control of the operation. A House resolution to endorse this idea failed in February 1832 by a vote of fifty-seven to sixty-seven. The vote reinforced a trend in the General Assembly that would persist throughout the antebellum era and ultimately result in the downfall of public improvements in western Virginia. Delegates voting in favor of public control of the statewide canal represented Virginia's aspiring commercial centers, such as Petersburg, Richmond, and Norfolk, the counties along the route of the improvement, and western counties in general. Given that internal improvement votes often pitted those who stood to benefit from proximity to the work against those who did not, it is surprising that ten of the eleven delegates from northwestern Virginia, which did not stand to benefit directly from the James-Kanawha water route, voted in favor of state control. The issue therefore assumed a sectional character beyond immediate economic interest. Of the sixty-seven votes against state control of the statewide canal, only five came from delegates west of the Blue Ridge. The overwhelming majority of negative votes came from delegates representing the agricultural tidewater and piedmont counties. The sectional character of the vote suggests that eastern rural slaveholding interests voted against state control and urban and western interests voted for it. Had legislative apportionment represented the distribution of Virginia's free population, instead of a mixture of free and slave, the issue might have turned out differently. The dominance of eastern slaveholding interests in the General Assembly, however, ruled out public assumption of the James River and Kanawha project.[16]

As sectional squabbling intensified and coal interests in the Kanawha Valley fumed, the James River Company underwent yet another transformation in the early 1830s. The legislature chartered a completely new firm in 1832, the James River and Kanawha Company (JRKC), and used the commonwealth's shares of the James River Company as an immediate transfusion of capital. Private investors provided the remaining three-fifths of the JRKC's subscriptions in 1835. Unfortunately, western Virginians saw little to cheer about in the policy of the JRKC, as they had in its previous manifestations. Work on the Kanawha River stalled due to poor management and inadequate funding, and, whenever western delegates took up the issue in the General Assembly, they could not secure more money for the JRKC to complete the task. As a result, colliers along the Kanawha remained shut out of the coal trade of the Ohio and Mississippi valleys.[17]

Virginia's floundering internal improvements program stifled the Kanawha Valley's attempts to reach new markets. The region's considerable production, which reached 133,300 tons by 1835, remained available only to local salt manufacturers. While the latter were important consumers of the region's coal, Kanawha Valley colliers hoped to expand the coal trade outside their state's boundaries. The 1840s saw a dramatic upswing in the consumption of coal along the Ohio and Mississippi, as traffic on the Ohio tripled in five years, from 106,414 tons in 1845 to 342,407 tons in 1850.[18] Kanawha County residents petitioning the JRKC for more work on their river remarked on "the almost magic growth of states and cities west of us," and they reminded the directors that "our valley is every year more and more a supply country to Cincinnati, Louisville, and the numerous towns and cities upon the Ohio and Mississippi."[19]

By the late 1840s, moreover, residents along the Kanawha could tout the discoveries of large deposits of cannel coal east of Charleston as yet another example of the region's mineral wealth. Cannel coal, a subclassification of bituminous coal which is easily rendered into liquid through a crushing and distilling process, was used in the manufacture of both lubricating and illuminating oils in the mid-nineteenth century. This discovery boosted the hopes of Kanawha Valley residents even higher, and a flood of new firms chartered to manufacture coal oil in the region. The demand for illuminating oil in eastern cities, local citizens argued to the General Assembly, would establish a profitable eastward coal trade as well. But the Kanawha Valley remained without water access to the eastern terminus of Virginia's statewide canal at Buchanan, as that portion of the James-Kanawha route had never materialized. Thus, they claimed, "we are completely cut off from all intercourse and trade with her interior counties, as if we were beyond the Rocky mountains."[20]

There are a number of possible explanations for the new firm's inability to improve the Kanawha River to the satisfaction of the local coal trade during the 1840s. Perhaps the directors planned to work on the Kanawha River improvement, but the financial crises of the late 1830s made raising money difficult for the venture. The JRKC faced mounting opposition from the advocates of a statewide railway, so the emergence of railroads may also have drawn needed capital away from the statewide project. Critics of the project often charged the management of the JRKC with squandering the aid of the Board of Public Works. Indeed, the firm's spending habits triggered two major legislative investigations during the 1840s to examine the potential mismanagement of state funds. The JRKC escaped from these investigations without any charges of malfeasance or charter violations, but the accusations of fraud and mismanagement damaged the firm's reputation.[21]

Yet the financial explanation makes little sense when one considers the special relationship between the JRKC and the Commonwealth of Virginia. A dearth of investment capital for internal improvements certainly existed in Virginia during the antebellum era, but the state's special relationship with the JRKC ensured that the firm would not go away quietly. Furthermore, if parsimony were the most important factor in the General Assembly's relationship with the JRKC, why did it authorize loans in excess of $1.5 million to the firm in 1846 to cover the costs of completing the canal to Buchanan and to connect the canal to tidewater via the Richmond dock?[22] The James River and Kanawha Company spent most of its existence in heavy debt, which certainly drew the ire of fiscal conservatives in the General Assembly. But it also enjoyed the sympathy of many legislators, who ensured that the firm remained solvent with state funds throughout the antebellum era. Opposition in the General Assembly blocked attempts to fund the JRKC fully, but it never succeeded in completely eliminating the state's role in the project.[23]

Another possibility is that party politics doomed the Kanawha improvement's interests within both the General Assembly and the Board of Public Works. After all, Kanawha County remained a stronghold of Whiggery for most of the antebellum era. Since the General Assembly was under Democratic control throughout this period, any partisan issues at work in the JRKC, the General Assembly, or the Board of Public Works would have hurt the interests of the Kanawha region. But this explanation also seems unlikely. First, votes on internal improvements in antebellum legislatures almost always assumed a sectional, not partisan, character. Although the two-party system that emerged in the 1830s and 1840s produced many hotly contested issues, internal improvements tended to provoke few party line battles.[24] Virginia's two-party system also was in a state of flux during these years. Throughout much of the 1830s Virginia's Democratic and Whig parties scrambled to establish distinctive positions on important political issues, and a clear divergence in partisan issues did not emerge in the Old Dominion until William Henry Harrison's "Log Cabin" presidential campaign of 1840. Thus, votes on the JRKC's Kanawha improvement occurred before any clear partisan discipline had emerged in the General Assembly.[25]

Instead, the main reason for the failure of the Kanawha improvement can be laid at the feet of eastern legislators, who formed a stiff opposition to the interests of western colliers. Despite constant petitioning by both the JRKC and citizens from western counties, funding for the Kanawha improvement could not pass the eastern-dominated legislature. Because alternative railroad routes such as the Baltimore and Ohio, Staunton and Potomac, and Lynchburg and New River railroads also failed to pass the eastern-dominated legislature, no relief was in sight for western colliers. In 1845, for example, a group of stockholders of the JRKC abandoned the idea of an all-water route and instead hoped to build a railroad to connect Charleston with the canal at Buchanan and then to improve the Kanawha from Charleston to the Ohio River. The petition drafted by the stockholders asked for funds specifically earmarked for the Kanawha River improvement, but it was voted down in a sectional vote. In 1848 the JRKC's chief engineer urged "upon the attention of the legislature" that the Kanawha improvement required immediate funding. "If developments of mineral wealth, which almost exceed the limits of computation, and an agricultural region unsurpassed in fertility by any portion of the state, form just claims to consideration," he wrote, "the appeals of the patriotic citizens of Kanawha ought not and cannot longer go unheeded and disregarded." The JRKC's board concurred with its engineer and specifically asked for $600,000 from the General Assembly to construct a lock and dam navigation that would make the Kanawha navigable for steamboats and open "the wealth of their numerous mines." This issue also failed in the General Assembly. By 1850 the commonwealth estimated that the JRKC had expended $5.16 million of state funds for the completion of their improvement, but only about 3 percent made it to the Kanawha region.[26]

Because sectional power in the legislature depended so critically upon apportionment, it is not surprising that Kanawha County emerged as one of the centers of the reform movement to reapportion the Virginia legislature at the same time that its coal trade suffered from the actions of an eastern-dominated legislature. Sectional rivalries in Virginia became especially tense after a raucous debate in the General Assembly over the future of slavery in the Old Dominion during the session of 1831–32. At that time antislavery forces nearly passed a provision for gradual emancipation. Following this close shave with emancipation, many large slaveholders of the east considered western reformers second only to New England abolitionists in threatening the future of the "peculiar institution" in Virginia. Western reformers vehemently denied that they would ever free slaves or grant free blacks equality; all they sought was equality among white Virginians. In 1841 Benjamin H. Smith, George Summers, and Spicer Patrick, all of whom had represented Kanawha County in the General Assembly, headlined a lengthy petition to allow for universal white male suffrage and to reorganize the legislature. "A large and decided majority of delegates and senators in the east has been insisted on as essential to the safety of the slave owners in that quarter," they claimed, "and great efforts were made to alarm the holders of that species of property, with the dangers that might arise from western influence in legislation." Petitioners argued that, because Kanawha County contained 2,560 slaves in 1840, or nearly 20 percent of its population, eastern delegates had no reason to suspect that all westerners were in league with abolitionists. "It is utterly impossible to extend equal rights to different races of men," the Kanawha petitioners maintained, and, because "the political community consists alone of the white population," their interests could not be intertwined with those of northern abolitionists.[27]

Yet slavery does not explain all of the sectional tension during this period. Eastern legislators might accept the argument that westerners could be trusted to preserve slavery. But they could not stomach the attempts by westerners to shift tax revenues raised mainly in the east into developmental projects in the west. Reapportionment also fell along fiscal lines. In 1842 the House commit-

tee selected to comment on this appeal concurred that Virginia should reapportion its legislature to reflect "a just ratio in reference to white population"
and stated that the main purpose behind the three-fifths compromise in the
1830 convention was "to secure a sectional ascendancy which could not otherwise be preserved; and that indeed is openly avowed." The eastern delegates
who served on the House committee issued their own minority report, in
which they argued for the preservation of the General Assembly's apportionment because realignment would threaten their rights of property. After all,
the easterners argued, "government is intended for the protection of *persons*
and *property*, and its form must be moulded as interest and necessity require."
Undoubtedly, a great deal of this anxiety revolved around the redistribution
of tax dollars to benefit western projects such as the Kanawha improvement
and the future of human chattel in Virginia. In order to preserve their property interests, eastern delegates argued, "that as we have *equal* rights of persons, and *greater* rights of property, and some of those rights in a measure peculiar to our society, we have a greater interest in the common stake, and ought
to possess an authority in proportion to that interest, and adequate to its protection." "It is not intended to affirm that the right of suffrage in the *individual* ought to be in proportion to his property," they informed their western
colleagues, "but that where there are different and distinct interests existing in
masses sufficiently large to form important objects of government, the rule
applies."[28]

Industrial development in the west and more equitable reapportionment
thus went hand in hand, and the linkages between political reform and internal improvements in antebellum Virginia were hardly circumstantial. The fortunes of the Kanawha region's coal trade, in fact, became intertwined with the
cause of reapportionment in the antebellum era. Many of the same community leaders who agitated for the white basis of representation also signed on
to the cause of the JRKC.[29] In both causes westerners argued that justice and
patriotism were on their side. Rather than advancing the interests of one section over another, most western delegates agreed, white representation, like
the completion of the JRKC, would in fact unite the Old Dominion. "If Virginians are tied down to a narrow selfish policy, that looks only to their immediate sectional interests," a western petition in favor of more state funding

for the JRKC argued, "the state cannot keep pace with her more enterprising neighbors." Given that many western Virginians owned slaves that were vital to the operation of the salt and coal trade, reformers of the Kanawha region questioned the justice of slave representation, not the legitimacy of slavery itself. This important distinction, however, fell upon deaf ears in Richmond.[30]

Constitutional reforms enacted in 1851 relieved some of the sectional tension in Virginia by extending the vote to all white men and providing for a reorganization of the General Assembly in the 1860s. The political changes of the 1850s never quite dispelled the belief among eastern delegates representing rural counties that the manufacturing interests of western Virginia sought to undermine slave-based agriculture in the Old Dominion or the conviction among white westerners that their eastern brethren considered the preservation of slavery more important than the maintenance of democratic institutions. Without the improved navigation, coal and salt exports gave way to competitors from farther up the Ohio River. Consequently, coal production in once promising Kanawha County significantly dropped during the 1850s. Figure 3.1 illustrates the steady rise and relatively rapid decline of production in the Kanawha basin.

In fact, the James River and Kanawha Company's neglect of the Kanawha River improvement served as a major stumbling block to western Virginia's coal trade up to the eve of the Civil War. A few privately chartered firms attempted to develop the cannel coal trade yet with little success. Mining and manufacturing companies with the authority both to mine and to transport coal also appeared in the Kanawha Valley during the 1850s but found themselves limited by the JRKC's influence in the region. Many of these companies found the Kanawha navigation itself to be the major problem. In 1857 the Old Dominion Coal and Iron Mining and Manufacturing Company lost ten thousand bushels of coal when one of its flatboats struck a rock and sank in the Kanawha River. The JRKC's sluggish action on the Kanawha frustrated the efforts of local colliers to raise capital and attract investors. In the summer of 1857, for example, John Barry was in New York City raising money for Kanawha County's Paint Creek Coal Company. The great potential of the Kanawha River coal trade constituted a major selling point of Barry's employers. But, when Barry heard the news that the Kanawha River improvement

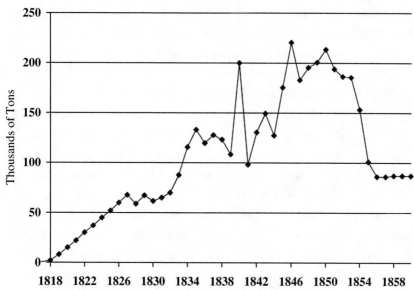

Fig. 3.1. Bituminous coal production of Kanawha County, 1818–1860.
(Eavenson, *First Century and a Quarter of American Coal Industry,* 507–8.)

was a "dead Lock waiting for further action of the Legislature," he regretfully
noted to the collier Christopher Quarles Tompkins that "this I presume is
equivalent to the PCCo [Paint Creek Coal Company] remaining status quo."
Efforts to raise much needed capital failed, and a year later Barry confided to
Tompkins that "nothing can be accomplished in the Paint Creek region until
the River improvement is secure."[31]

The final chapter in the history of the antebellum Kanawha improvement
is, not surprisingly, one of discord, mistrust, and failed opportunities. In 1858
the General Assembly responded to pleas for aid by setting up a separate
Kanawha Board staffed by local residents to oversee future improvements to
the Kanawha River improvement. Two of the five members of the Kanawha
Board were agents of new coal companies in the region, and their scheme to
float a loan to restart work on the river signaled a new era of action for many
Kanawha region colliers. The James River and Kanawha Company responded
in 1859 by attempting to block the loan and denounced the Kanawha Board's
authority to supervise construction on their property. In the end the General
Assembly supported the Kanawha Board and authorized $300,000 of the

state's stock in the JRKC to be used solely for the improvement of the Kanawha. Before extensive work could be completed on the river, however, the advent of the Civil War settled the question of Virginia's role in the Kanawha improvement permanently.[32]

State funding of the Old Dominion's centerpiece internal improvements network favored the established tidewater and piedmont regions over the budding industrial interests of the state. This policy, made possible by an unequal apportionment of the General Assembly, favored a conservative vision of the Old Dominion's state government. Amid this sectional controversy in Virginia politics, the coal trade at the western end of the James River and Kanawha route suffered. Colliers in Virginia's western counties found themselves shut out of the valuable Ohio-Mississippi valley markets of the antebellum era. One could argue that the Kanawha improvement was not an economic panacea, as western Virginians claimed. What is clear, however, is that conservative ideas about the role of the state in economic development suffered at the hands of sectional politics.

A zero-sum game continued to run in Virginia's antebellum General Assembly, in which the gain of one section entailed a corresponding loss for the other. Since many eastern delegates viewed the growth of western Virginia as a threat to their financial well-being, they used their preponderance of votes to block western attempts at internal improvements. Some of their mistrust can be attributed to concerns over the future of slavery, but the sectional squabbles of antebellum Virginia owed as much to different ideas about "development" as they did to discussions of slavery. Virginians east of the Alleghenies feared that the enhanced economic prestige of the western counties would, at the very least, lead to greater spending on improvements beyond their immediate interest and increases in the taxation of slaves. Although proslavery politicians from the west such as Henry Wise helped erase any doubts about the future of slavery in Virginia, sectional animosity persisted even after Virginia's Reform Constitution of 1850–51. Eastern attempts to undermine industrial endeavors such as the Kanawha coal trade in Virginia did not, therefore, reflect sectional anxiety over slavery as much as they signified an ongoing struggle between conservative and democratic visions for the future of Virginia's polity.[33]

MAIN LINE, BRANCH LINE, OR BOTH?

THE PENNSYLVANIA STATE WORKS

At first glance Pennsylvania's geography appeared likely to trigger a similar
sectional conflict to that which afflicted Virginia's public works program. After all, Pennsylvania's mountains divided the state into at least two, if not three,
distinct sections with different market orientations: Philadelphia, looking to
the eastern seaboard trade; Pittsburgh, eyeing the growing commerce of the
Ohio Valley; and much of central Pennsylvania, seeking to link up with New
York City to the north and Baltimore to the south. Pennsylvania's responsive
state government, however, made a critical difference in mediating these interests in the construction of a statewide internal improvements network.
Unlike tidewater and piedmont delegates in Virginia, who wielded political
power in the General Assembly at the expense of western interests, the representatives of Pennsylvania's competing regional interests, more evenly balanced in the legislature, compromised to create a more widespread, albeit
bloated and inefficient, public works program. In terms of the coal trade the
Pennsylvania State Works linked yet undeveloped coalfields to urban markets,
increased coal traffic across the state, and generally avoided competing with
the already established anthracite canals and railroads of eastern Pennsylvania. The State Works thus stood as an example of the strengths and weaknesses
of the legislature's flexible policy making: strengths in its ability to expand the
state's coal trade and weaknesses in the way that the ambitious program nearly
bankrupted the state.

Pennsylvania's first publicly funded internal improvements system unfolded in the same spirit as Virginia's James River and Kanawha route. During the late eighteenth and early nineteenth centuries the state government
had played a limited role in transportation projects. Since 1785, when the legislature passed an act to improve the Chambersburg-Pittsburgh turnpike,
Pennsylvania had promoted internal improvements mainly by chartering private concerns and occasionally by subscribing to their stock. By the 1820s
Pennsylvania had one of the most extensive turnpike and canal systems in the
nation, built largely through private initiative and a blend of public and private investment. Of the $6.4 million subscribed and appropriated in turnpike
capital by 1822, about $1.8 million came from Pennsylvania's state treasury. A

year later nearly $10.5 million had been invested in Pennsylvania internal improvements, with over $2.3 million of that amount subsidized by public funds.[34]

New York's experience with the Erie Canal, however, proved that coordinating large-scale projects was essential in order to exploit western markets. After its water link with the Great Lakes system was completed in October 1825, New York City threatened to dominate both eastern and western trade in the United States—much to the dismay of many Philadelphia merchants. To the south Maryland's aggressive move to connect Baltimore with the Ohio River by turnpike and canal also jeopardized the commercial interests of Philadelphia. Legislators in Harrisburg responded to the early success of the Erie Canal by passing a bill in March 1824 which provided for the appointment of three commissioners to explore potential routes from Pittsburgh to Philadelphia. This commission's report, completed in February 1825, recommended the immediate construction of a canal from Pittsburgh to Philadelphia. Revenues from this project would, in the words of the committee, "support the government and educate every child in the commonwealth" as well as provide an impetus to the economic development of the state. The report thus laid the foundation for a statewide system of internal improvements in Pennsylvania modeled on the Erie Canal system which would become known simply as Pennsylvania's "State Works."[35]

The political birth of the State Works occurred in 1826, but the deliberations over the final shape of the project dragged on for another two years. Organizations such as the Society for the Promotion of Internal Improvements provided a strong theoretical argument for the construction of a trunk line across the state, but the structure of Pennsylvania's legislature gave both voice and power to dissenting legislators who wanted the State Works to benefit their districts also. Over the course of this struggle, partisan affiliations gave way to sectional loyalties, as representatives fought bitterly for the inclusion of their districts in the State Works. Counties in southern Pennsylvania disliked the idea of a main line because it would draw traffic away from the Chesapeake and Ohio Canal, which planned to build a trunk line through their area. The residents of the upper Susquehanna Valley, which included a number of coal interests along the river's northern branch, also argued that the plan, as it stood in 1826, ignored the rich mineral resources of their region. These interests dovetailed with the more general grievances from counties outside the reach of the proposed main line to create a solid bloc of votes in the legislature against the project.[36]

Unlike in Virginia, where sectional animosity effectively vetoed statewide improvements, Pennsylvania's political conflict over the shape of its internal improvements system resulted in the addition of branch line canals to the State Works. Branch lines linked many of the areas isolated from the east-west traffic of the Main Line and quelled, at least temporarily, major opposition to the construction of the State Works. During the late 1820s and early 1830s legislators separated into two main camps on transportation issues. The "Main Liners" viewed the trunk line connecting the western and eastern extremities of the state as most important, while "Branch Liners" saw intrastate linkages as the real value of state-sponsored improvements. Both sides, however, agreed on the principle of state-supported internal improvements and on the benefits such projects promised for the commonwealth's economic future. For this reason they had to hammer out a compromise in which the canal commissioners would construct branch canals simultaneously with the Main Line. Of course, these side projects diverted important financial and organizational resources from the Main Line and delayed its completion, but they were absolutely necessary to secure sufficient legislative support for the trunk line in Harrisburg. "Hence, to secure a majority in favour of the main object," observed strong Main Liner Mathew Carey in 1831, "it was imperiously necessary to provide for a number of minor ones."[37]

The blend of branch and main line construction in the State Works constitutes a prime example of a legislative process known as "logrolling," in which disparate interests support their respective projects in a single bill in order to maximize support and ensure its passage. Single legislative acts appeared regularly in the Pennsylvania legislature which covered internal improvement funding, divorces, pensions, turnpike stock subscriptions, and a host of other topics seemingly strung together at random. These "omnibus bills" often made it difficult for members to vote either for or against a single piece of legislation, thus shifting the policy implications of a bill from voting to its initial construction. Pennsylvania's reform constitution of 1838 effectively blocked formal logrolling by restricting each legislative act to only one subject. Yet informal logrolling, in which legislators promise to vote for one another's projects, remains a common feature in legislative politics even today. For this reason many contemporaries and historians targeted logrolling as a major reason for government waste and corruption in internal improvement programs in all states. Such practices undermined the efficiency of the Pennsylvania State Works at its very inception and doomed the project to financial failure.[38]

Undoubtedly, the logrolling evident in the State Works created problems for the speedy completion of the Main Line and more than likely resulted in the support of a few branch projects of dubious merit. Yet, when compared to the paralysis caused by sectional animosity in Virginia during this period, the Pennsylvania State Works must be viewed as a political success. Unlike Virginians, Pennsylvanians periodically reapportioned their legislature in order to reflect population shifts and guaranteed the vote to virtually all white residents throughout the antebellum era. Equity in representation made compromise a necessity in the Pennsylvania legislature, as no single region or interest could garner enough votes to ram an issue down the rest of the House or Senate's throat. Representatives often approached this decentralized policy-making structure with a spirit of egalitarianism which further complicated matters; an internal improvement in their view was not a privilege but a right. The resulting pattern of widely dispersed systems became common among northern states during the canal boom of the 1830s.[39]

Instead of serving as a barrier to development, logrolling promoted it by transforming the State Works from a contentious and divisive issue on the floor of Pennsylvania's House and Senate into a real program. Granted, this process owed as much to cloakroom and boardinghouse negotiations as it did the more formal avenues of policy making, but such was the stuff of democratic politics. After all, didn't the eventual shape of the State Works serve both the Main Liners and Branch Liners? Supporters of internal improvement such as Mathew Carey argued that what is "vulgarly called '*log-rolling*,' [was] the result of a spirit of compromise; and which, unless when employed for the attainment of improper or unjust objects, or dictated by a spirit of faction or intrigue, is often not only necessary, but even perfectly justifiable." The sacrifice of economy and speed in the construction of the Main Line in order to build more branch lines—a decision made in part through considerations of Pennsylvania's coal trade—ultimately saved the political future of the State Works.[40]

The coal trade often appeared in discussions of the State Works. Advocates of both the Main Line and the branch canals used the example of the anthracite canals as justification for the massive system. For example, an 1830 report of the House Ways and Means Committee cited the Schuylkill Navigation Company's success with coal as proof of the financial soundness of a state-operated system and referred to the "exhaustless stores of mineral wealth" in Pennsylvania. "The public disbursements on the canals have not been buried in the earth," the committee reminded its colleagues, "nor have they taken to themselves wings and flown away." Supporters also used the example of the Schuylkill and Lehigh canals to highlight the synergetic development between mining regions and internal improvements. This line of argument became especially useful for the canal commissioners when critics pointed out the lack of immediate remuneration to the State Works.[41]

From its earliest conception promoters of the State Works also invoked the example of Great Britain and envisioned their state taking a similar path of development. They argued that branch canals carrying coal as their primary traffic would provide early business for the main line and help promote the growth of iron, salt, and other industries all along the State Works. Just as Great Britain built its economic prowess on coal, supporters of public improvements claimed, so could Pennsylvania. New York's lack of coal reserves made this patriotic argument even more appealing to policy makers concerned about economic competition from the Empire State. Mathew Carey calculated that in Britain per capita consumption was nearly a ton for every individual. His expectation that Pennsylvanians would match that level presented a "delightful prospect for Pennsylvania canals" in terms of future toll revenues. "The immense advantages which the coal treasures of this state present to the mind, in calculating the extent of future consumption," crowed Carey, "are of the most exhilarating character."[42]

By 1835 the State Works began to take shape. That year workers completed over 350 miles of the Main Line between Pittsburgh and Philadelphia at an estimated cost of $12 million.[43] Although known primarily as a canal, the trunk line of the State Works was actually a blend of land and water navigation. From east to west the trunk line began with a railroad link from Philadelphia to Columbia on the Susquehanna River. From there the Eastern and Juniata divisions followed the Susquehanna River eastward to the Juniata River Valley, where the Main Line terminated at Hollidaysburg in mountainous central Pennsylvania. The Allegheny Portage Railroad, an ingenious system of stationary engines pulling cars on inclined planes, provided a link from Hollidaysburg to Johnston. Finally, the Western division of the canal completed the 104-mile journey from Johnstown to Pittsburgh along various rivers until it ended at the Monongahela River. The Main Line passed through Pennsylvania's western bituminous region along the line of the Allegheny River, northeast of Pittsburgh, and also passed just north of the Broad Top bituminous coalfield in the south-central portion of the state.[44]

Of the ten branch, or lateral, additions to the State Works, three stand out in the state's coal trade, but only one seemed to draw any interest from private competitors. First, the North Branch division (completed in 1834), which ran along the North Branch of the Susquehanna River and ended at Lackawanna Creek, linked the Main Line to the northern anthracite field via the Susquehanna River system. The West Branch division (1835) extended the navigational capacities of the Susquehanna River to the western bituminous fields of north-central Pennsylvania. Finally, the Delaware division (1832) expanded the traffic of that river from Easton, the terminus of the Lehigh Coal and Navigation Company, to Bristol, a town eighteen miles north of Philadelphia. The first two lines linked remote areas of Pennsylvania to market centers—thus reflecting the political character of the State Works. The Delaware division allowed the state to tap into the growing Lehigh River traffic, much to the chagrin of the Lehigh Coal and Navigation Company. The potential competition for the LCNC provided by the State Works undoubtedly pleased rival coal mining interests in the Schuylkill Region, and supporters of the Delaware division tapped into popular anti-charter rhetoric to justify their project's redundancy with the LCNC's river improvement. Since the legislature already had bestowed the LCNC with the power to both mine and ship coal, Delaware division supporters argued, then why not provide colliers in the Lehigh Region with an alternative route to market?[45]

The birth of the State Works stood as a testament to the necessity of Pennsylvania's legislature to compromise between disparate economic interests, but its early childhood hardly prospered as a result of the system created by logrolling and other political tactics. For one thing, the State Works suffered from the lack of firm administration. The Board of Canal Commissioners, which consisted of three political appointees, supervised the entire system in theory. But, before the State Works were completed, legislators recognized that they had constructed an institution simultaneously susceptible to faulty administration and difficult to reform. In the words of one 1833 committee, "even though no corrupt influence be exerted," the lack of a hierarchical structure of administration made it "extremely difficult to detect whatever defects may exist in the operation of such an engine." Complaints about the distribution of contracts, methods of payment, and other abuses of the system plagued the State Works throughout its existence. More often than not, legislators from the minority party brought these problems to the public light. Unlike the struggle between Main Liners and Branch Liners, partisanship in this instance exacted its toll upon the State Works. Rather than radically restructure the administrative system to ensure that it was more responsible and less susceptible to corruption, legislative critics sought instead to replace the other party's villains with villains of their own. Thus, the administration of the State Works remained hopelessly inadequate throughout its entire existence.[46]

The political nature of the program demonstrated the absurd lengths to which Pennsylvania's state government would go to promote its own coal industry. When the canal commissioners reported that the state-owned Philadelphia and Columbia Railroad burned over $6,000 worth of wood in its locomotives and less than $1,000 worth of coke and bituminous coal in 1835, anthracite interests went to work. By 1838 Governor Joseph Ritner announced the "complete success of the trials to use anthracite coal as a fuel for generating steam in the locomotive engines," and the Philadelphia and Columbia purchased several coal-burning engines. Despite the political popularity of anthracite engines, by 1845 the Philadelphia and Columbia found anthracite engines impractical for its lines and went back to burning wood and bituminous coal. Design flaws and efficiency problems killed the politically attractive idea of burning anthracite coal in the locomotive engines of the State

Works but not before the state expended thousands of dollars on this idea. The privately owned Philadelphia and Reading implemented anthracite-burning engines on its line, but this transition did not occur until the mid-1850s.[47]

For the State Works events such as the anthracite steam engine debacle allowed the fissure between political expediency and financial solvency to widen quickly into a chasm. Soon after the opening of the main line, the Panic of 1837 made further loans and credit arrangements to pay for the State Works difficult to secure. In the face of the financial crunch facing their constituents, state legislators in Harrisburg steadfastly refused to raise taxes significantly (one of them a tax on coal) and only passed politically safe measures, such as an 1840 law establishing a light tax on dividend-paying stock, household furniture, luxury goods, and the salaries of state employees. Eventually, the legislature found it could not avoid this fiscal crisis. In 1842 the commonwealth suspended interest payments to its creditors, as it could no longer meet them. During the following session members of the Committee of Ways and Means reported that, in the face of Pennsylvania's funded debt of $38 million, "those palmy days of *apparent* prosperity have passed, and we are now thrown upon our own resources." The frightening possibility of statewide bankruptcy overshadowed the political consequences of taxation. In 1844 Governor David Rittenhouse Porter initiated a more radical program of fiscal responsibility in which the commonwealth raised taxes on personal and real estate in order to "furnish the necessary amount to discharge the interest upon the public debt, and thus ensure the fidelity of the state to her engagements."[48]

As a result of the program's runaway debt and the radical solutions needed to control it, public support for the State Works also waned. In 1840 the canal commissioners admitted in their annual report that in the infancy of the system "many works were commenced from which little immediate benefit is to be expected," and that some of the State Works were "constructed, not upon the soundest and most economical plan." In order to save the sagging reputation of the State Works, however, they cited "innumerable coal and iron regions now untouched" that future canal projects would serve, as well as the general benefits the system would have to agriculture and industry. That same session, members of the Committee on Inland Navigation and Internal Improvement recommended freezing the "mad career" of the State Works to allow Pennsylvania "a short period of repose, in order to enable her to restore the deranged state of her finances to a healthy condition." In a complete reversal of the rhetoric that characterized the early period of the State Works, the members argued that patriotic citizens "will assent cheerfully to a suspension of those works for a short period, to enable their favorite Commonwealth to extricate herself from the labyrinth in which she finds herself so deeply involved."[49]

The sheer size of the public debt and the lack of immediate remuneration of the State Works thus undermined its popularity as a public venture, and the legislature contemplated liquidating it as a means of restoring faith in the credit of Pennsylvania and relieving the commonwealth of the financial burden of further construction and upkeep. The sale of the main line connecting Philadelphia and Pittsburgh, the crown jewel of the State Works, evolved into a lengthy and controversial process. In 1844 the question of selling the Main Line passed a popular referendum by a margin of over twenty thousand votes. That same year an act passed which authorized the sale of the Philadelphia-to-Pittsburgh route for twenty million dollars, a price that no private concern was willing to pay for a line of navigation which cost sixteen million dollars to build. The Pennsylvania Railroad, chartered in 1846, eventually emerged as a prospective buyer. Its line paralleled the Main Line, more or less, and the Pennsylvania Railroad's directors resented the fact that the firm paid a tonnage tax to the state treasury because it competed with the State Works for traffic.[50]

In 1854 the commonwealth lowered its asking price to $10 million dollars, and in 1857 it accepted the offer of the Pennsylvania Railroad Company to buy the Main Line for $7.5 million. The directors of the Pennsylvania Railroad also agreed to pay a $2.5 million dollar bonus to release the company from its tonnage tax, which was later declared unconstitutional by Pennsylvania's Supreme Court. By this time the patriotic zeal of the 1820s and 1830s had dwindled to a faint echo, and the tone on the floor of the Pennsylvania legislature was much less benign toward public investment in transportation. "It may safely be assumed that the people of Pennsylvania have no faith in the assertion so often made, but not yet verified, that any part of the public improvements are of sufficient value, present or prospective, to be retained by the State," argued the House Committee of Ways and Means in 1858. The com-

mittee went on to conclude that internal improvements "cannot be successfully or profitably managed by the State; that their retention is more likely to
prove an annual loss than a source of profit to the Treasury; and that a proper
regard for the public welfare and a decent respect for the wishes of the people, demand that they should be sold without unnecessary delay."[51]

Despite its cruel political demise, Pennsylvania's State Works proved helpful in developing some of the state's most valuable coalfields. The Western division of the Main Line, which was completed to Johnstown by 1830, facilitated the growth of the small-scale bituminous mining enterprises along the
Conemaugh and Allegheny rivers in Westmoreland, Indiana, and Armstrong
counties. The linkages between Pittsburgh and these rich bituminous fields
along the Western division strengthened that city's position as a major distribution center for the shipment of bituminous coal to the Ohio and Mississippi valleys. In 1834 colliers in western Pennsylvania raised 380,000 tons of
coal, a 35 percent increase from their production level of 1829, the last full year
prior to the construction of the Western division. That same year *Niles Register* estimated that $100,000 worth of coal was annually sent down the Ohio
River from Pittsburgh. This valuable trade owed much of its vitality to the new
regions opened by the State Works. Pittsburgh thus emerged as the central
port in a vibrant coal-exporting region, while Kanawha Valley colliers suffered
from their marginal presence in Virginia's internal improvements network.
Politics, not nature, allowed Pittsburgh's coal trade to flourish in the 1830s and
1840s.[52]

The opening of the new bituminous fields in central Pennsylvania is
yet another example of the impact the State Works had upon Pennsylvania's
coal trade. Once considered too remote for viable investment, these bituminous fields thrived after the completion of the Juniata division, which linked
the Susquehanna to Hollidaysburg in the mountains of modern-day Blair
County. Colliers in these areas could ship their product eastward along an all-
water route to Harrisburg and Philadelphia and, after 1840, could reach the
Chesapeake Bay via the Susquehanna-Tidewater Canal. Production in the immediate area of Blair County leaped from 4,000 tons in 1836 to 31,000 in 1839,
a 675 percent increase in three years. The most complete run of data regarding the coal traffic through Hollidaysburg suggests that shipments over the

State Works originating in central Pennsylvania increased from 16,251 tons in 1843 to 53,977 tons ten years later.[53]

Outside of the Main Line, branch lines provided much-needed linkages between coal-producing areas and market centers within Pennsylvania. One such project tapped into the rich northern, or Wyoming, anthracite field in order to develop that area. The North Branch division, completed in 1834, connected Northumberland on the Susquehanna to Old Forge Dam, about ten miles north of Wilkes-Barre. The canal commissioners placed a price ceiling on toll rates so as to favor these relatively distant regions and spur the anthracite trade of the Susquehanna.[54] Consequently, more and more anthracite coal passed through the North Branch division, which both increased the coal mining around the Wilkes-Barre region and the toll revenues for the North Branch. In fact, when compared with the other major carrier of anthracite in the northern anthracite field, the Delaware and Hudson Canal Company, the North Branch competed quite well with private firms. Table 3.1 compares the traffic and share of the northern field's total production shipped on the State Works at Berwick with that of the Delaware and Hudson at its shipping point of Honesdale. Because data concerning coal traffic on the State Works is fragmentary, a complete run of the coal traffic at Berwick for the years exists only during the years from 1844 to 1849. Nonetheless, the comparison demonstrates that public and private carriers both ably served the northern anthracite field during its development.

In the end, however, the upkeep on the North Branch proved too much.

Table 3.1. Public and Private Carriers in the Northern Anthracite Field

	1844	1845	1846	1847	1848	1849
North Branch						
Coal shipped (tons)	116,018	178,401	188,375	429,033	237,271	259,080
% of Northern Field	23.8	29.7	27.4	52.6	26.2	26.8
Delaware and Hudson						
Coal shipped (tons)	251,005	273,535	320,000	386,203	437,500	454,240
% of Northern Field	51.5	45.5	46.5	47.4	48.3	47.0

Source: Eavenson, *First Century and a Quarter of American Coal Industry,* 496; Jones, *Anthracite-Tidewater Canals,* 103; Bishop, "State Works of Pennsylvania," 282.

The canal commissioners could not fend off competition from the Lehigh Coal and Navigation Company's rail link to Wilkes-Barre. The extension of the Delaware and Hudson's canal and railroad links into the northern anthracite field during the 1840s also bit into the North Branch's traffic. Along with the Main Line, the North Branch division was sold to private interests in 1857 and resurfaced as the North Branch Canal Company. The North Branch had earned a net revenue of nearly $1.3 million under state ownership, which marked it as one of the few lateral projects in which revenues exceeded the operating costs. Other branch projects were not so profitable under state ownership. For example, the net expenditures on the West Branch division exceeded revenues by $150,700 before it was sold in 1858. The Beaver and French Creek divisions lost $172,048 and $138,092, respectively, with the Beaver's deficit in revenues representing a third of the original cost. The state turned over control of these lateral canals after 1845, which makes their high losses all the more impressive. But the North Branch was among the most successful of the supplementary canals in the State Works for reasons other than the net return it earned on the commonwealth's investment. It succeeded in opening new anthracite coal regions to development and diverted a significant portion of the northern anthracite field's traffic through Pennsylvania.[55]

The other major branch canal affecting Pennsylvania's anthracite trade while under state control was the Delaware division. The Lehigh Coal and Navigation Company applied to the legislature for the right to improve the Delaware River during the 1820s but could not convince legislators that this privilege would prevent the LCNC from monopolizing traffic in northeastern Pennsylvania. Instead, the canal commissioners authorized their own survey along the Delaware in 1827 and began construction of the project that same year. In 1832 the commonwealth completed the canal that linked the southern terminus of the LCNC's canal at Easton with Bristol on the Delaware River. Although the canal commissioners intended the Delaware division to be an extension of the LCNC's system, problems with capacity became immediately evident. The LCNC's canal was built to accommodate boats of one hundred tons. The Delaware division was designed to handle boats carrying up to sixty-seven tons, but actual capacity in 1841 reached only fifty-five to sixty tons. The small capacity of the canal made the smooth transfer of coal barges from the

LCNC's system to the state canal at Easton difficult, if not impossible, which raised the cost of transporting Lehigh coal to the Philadelphia over the public canal. Furthermore, the Delaware division suffered from problems with its water supply. Spring freshets often damaged the works, and in dry seasons the canal could not even maintain five feet of water.[56]

But the faulty engineering on the Delaware division only accounted for a portion of the branch canal's problems. The political struggle between New York, Pennsylvania, and New Jersey for the Delaware division's traffic really doomed this venture to failure. Pennsylvania's canal commissioners remained wary of any actions that might benefit traffic flowing to neighboring states and attempted to keep the Delaware division advantageous to intrastate traffic. For example, the commissioners refused to construct an outlet lock at Black's Eddy to link the Delaware division with New Jersey's Delaware and Raritan Canal. Instead, coal shipments designated for New Jersey had to travel the length of the Delaware division to Bristol. The barges would then backtrack up the Delaware River to reach the Delaware and Raritan. Even when the commissioners agreed to the outlet lock in 1846, they still charged the same toll as if the boats had traveled to Bristol. The canal commissioners had hoped that they could profitably tap into the Lehigh coal trade, but the Delaware division failed miserably.[57]

Some unopened branch canals proved to be profitable routes for the coal trade after the commonwealth underwrote survey and initial construction costs, thus demonstrating the role that the State Works played in Pennsylvania's coal industry long after its ruin. For example, the Erie Extension Canal, which was to link the Shenango River with Lake Erie over a distance of roughly 106 miles, cost the state over three million dollars to build. When several petitions circulated in 1842 to abandon this expensive project, the Erie Extension's defenders cited the region's coal business "yet almost in embryo" and noted that further development will "in all probability disclose mines not now known." The boosters of the Erie Extension also presented detailed estimates of the high demand for coal in Buffalo and the surrounding region and promised that, with the need for steamboat fuel alone, "we shall be almost be overwhelmed with the magnitude of the commerce."[58] In the harsh climate of Pennsylvania's newfound fiscal responsibility, such predictions received a cold reception. In 1843 the commonwealth virtually gave away the Erie branch line to the Erie Canal Company, thus ending the public investment in the Lake Erie coal trade. During the late 1840s and early 1850s canal traffic in coal from Pittsburgh to Erie grew immensely. By 1853 Erie received 103,031 tons of bituminous coal and commanded a significant portion of the Great Lakes coal trade. In the long run investment in this branch canal was justified, but the short time horizon of legislative politics forced the State Works to liquidate the Erie Extension.[59]

There is no denying that the political origins of the State Works undermined the earning potential of state-funded internal improvements in Pennsylvania. The simultaneous construction of main and branch canals, the creative bookkeeping of agents and employees, and the lack of long-range planning combined to cripple the financial future of the State Works. In his landmark study of the public promotion of U.S. canals and railroads in the nineteenth century, Carter Goodrich argued that the driving force behind state-funded improvements was "developmental, and an assessment of its impact must take account of these diffused gains and their cumulative effect on the growth of the economy." The experience of Pennsylvania coal and the State Works demonstrated this axiom perfectly. State-funded canals linked undeveloped coal regions to market centers, added outlets to existing areas, and facilitated the transport of coal beyond Pennsylvania's boundaries. Without the immediate construction of branch and main lines spurred by the political environment at Harrisburg, moreover, the impact of state funds would have been considerably less.[60]

By the 1850s technological advances in fuel economy and steam power made the overland carriage of coal and other bulk commodities by railroad profitable. For the coal trade the rise of railroads presented a new challenge to both established and prospective mining concerns. In Virginia the political weight of the James River and Kanawha Canal—the centerpiece internal improvements project of Virginia through the 1850s—combined with other factors to frustrate the development of a statewide railroad system. In Pennsylvania the state government was largely divested from internal improvements by the time the railroad boom of the 1840s and 1850s hit the Keystone State.

Pennsylvania surely had its share of sectional rivalries, as railroads rose to challenge canals as major coal carriers, but the relocation of transportation policy from the legislature to the private sector in Pennsylvania avoided the zero-sum relationship played out in Richmond year after year. There the primacy of state funding in internal improvements had a devastating effect upon the development of Virginia's western bituminous coalfields, as sectional politics shaped the Old Dominion's transportation network.

In all fairness the corruption of the Pennsylvania State Works and the ways in which its faulty operation ran the state to the brink of financial ruin hardly served as a model of public enterprise. What is usually left out in discussions of the State Works, however, is that internal improvement projects rarely pay for themselves in the short run, and nearly all American state-funded transportation systems went through times of financial hardship followed by a political backlash. Furthermore, the beneficial effects of Pennsylvania's sprawling system—both in terms of competition for private lines and in linking markets at an early stage—cannot be quantified and therefore escapes the attention of many economic historians. Not that Pennsylvania's post–State Works transportation policy made much sense either. In 1857 Pottsville's *Miners' Journal* complained of "an entire lack of system in regulating the supply to the demand, conjoined with so many flimsy paper calculations, propagated to advance selfish interests," which converted the coal business into a game of chance in which "the longest pole knocks the persimmon." Order or rationality, it seems, did not characterize the private carriers that serviced the state's coal trade any more than they did the State Works.[61]

It cannot be denied, however, that Pennsylvania's more flexible and open governmental structure produced a transportation network that served both the short- and long-term interests of the coal trade. The brief, but unfortunate, history of the State Works serves as a testament to the brilliance and the tragedy of the Keystone State's flexible style of policy making; it is highly unlikely that Pennsylvania would have emerged as an economic powerhouse without its state-funded canals, but for a time it seemed equally uncertain that the state treasury would ever recover. The coal trade in both existing and undeveloped areas benefited a great deal from the logrolling and other political hijinks surrounding the Pennsylvania State Works.

Colliers in western Virginia suffered a different fate. The Old Dominion's experience with state-funded internal improvements reflected the shortcomings of its political economy. In order to maintain power in the General Assembly, delegates from counties with large slave populations actively suppressed attempts to reform Virginia's state structure. Western projects such as the Kanawha improvement fell victim to this protracted sectional struggle for power. The Kanawha Valley coal trade remained limited to fulfilling local demand, as colliers could not count upon the regular water route they needed in order to export coal to the Ohio and Mississippi valleys. As long as state funding remained subject to the political influence of Virginia's minority of conservative easterners, it was a foregone conclusion that western interests would remain on the margins of internal improvement policy in the Old Dominion.

"Hidden Treasures" AND Nasty Politics

Antebellum Geological Surveys in Pennsylvania and Virginia

*Y*ou can scarcely imagine how, beauty & deformity, art & nature, sweetness & filth, sterility & mud are strangely mixed in this singular place," a twenty-year-old J. Peter Lesley wrote to his sister while analyzing Pennsylvania's southern anthracite field in 1839. Although the flowers were "strangely contrasted with the mud & coal" in the mines outside of Pottsville, the young scientist described himself as "uncommonly spirited & happy in it." The anthracite trade had completely transformed this mountain town, as hardly a minute went by without Lesley hearing mention of "drifts," "shafts," "wharves," or "coal." He wrote to his grandmother: "I don't think that the trees have learnt to rustle or the birds to whistle 'coal' yet, tho' it would probably take a cool headed speculator to calculate how long it will be before that happens." Lesley's position as a representative of the state geological survey, however, made him a suspect character in the eyes of many local colliers. "We keep our business as secret as possible, such is the queer feeling often exhibited towards the institution & measures of the survey," Lesley wrote a few weeks later. "I have enter'd no society as yet, satisfied with my books, my horse, my pen & my fellow boarders."[1]

Lesley's stealth was wise. The rude state of geological knowledge, combined with the wishful optimism of mining entrepreneurs, obscured the location of profitable coal deposits throughout the first three decades of the nineteenth century and could turn a geologist from hero to goat in a matter of weeks. Stories of massive veins lying close to the surface in unexplored areas materialized—many from thin air—as the demand for coal rose. The speculative frenzy that occurred in Schuylkill County during the 1820s and 1830s, for example, set a very bad precedent. Many prospective miners of anthracite paid exorbitant prices for tracts of land that would never surrender coal. Others arrived to find themselves the victims of outright fraud. The *Philadelphia Gazette* reported in 1829 that enterprising lawyers had drawn up and sold enough land warrants to cover twice the surface area of Schuylkill County. The established Virginia coal industry also suffered from unscientific speculation. Exactly where and what type of coal existed in the Richmond basin remained largely up to the imagination of local boosters and owners of land. The promise of limitless supplies of coal hidden in the mountains and valleys of Pennsylvania and Virginia sustained interest in the coal trade, but the dubious claims of land speculators and local boosters frustrated prospective coal barons.[2]

For careful investors the only way to avoid this problem was to enlist the aid of an expert to survey the prospective territory. American scientists often served as consultants to such projects. Yale University's Benjamin Silliman, a chemist by training, traveled to Virginia in 1836 to assess the value of the Raymond Mining Company's land for a group of English buyers, and Lesley's later career as a geological consultant proved quite lucrative. But hiring a scientist to survey tracts was expensive and time-consuming, especially in the vast bituminous fields of both states' western counties. A less burdensome solution for entrepreneurs was to create a statewide geological survey, authorized and paid for by the state itself. The legislature would appoint a state geologist with a yearly stipend, who would then undertake an extensive exploration of the state's geology for a few years, summarize the findings in annual progress reports, and eventually publish an authoritative summary of the survey's work. North Carolina had pioneered the state geological survey in 1823, and South Carolina, Massachusetts, and Tennessee all boasted that they had similar programs in place by 1831. By 1850 twenty states had commissioned geological surveys.[3]

These antebellum state geological surveys became a blend of economic and scientific aspirations. Entrepreneurs found them attractive because they shifted the financial burden of exploration to the public sphere. Scientists saw state surveys as both steady employment and an opportunity to work in the field in the pursuit of knowledge. Politicians, however, were a tougher sell. Although many legislators sympathized with the idea of promoting their state's mineral resources, sponsoring an extensive geological survey diverted money that could just as easily be allocated elsewhere. In order to get around the skepticism of politicians and push their projects through reluctant state legislatures, supporters of early geological surveys employed patriotic rhetoric. The ideal antebellum survey combined economic, scientific, and patriotic agendas within the scope of a single institution.[4]

Both Virginia and Pennsylvania initiated surveys in the 1830s. Their state geologists, Henry Darwin Rogers and William Barton Rogers, respectively, were academic scientists with stellar credentials who also happened to be brothers. Despite the familial bonds of the two geologists, the execution of Virginia and Pennsylvania's surveys became profoundly influenced by their political context. Henry Darwin Rogers's Pennsylvania survey, which was originally promoted by Philadelphia's powerful academic community and undertaken with great zeal, suffered from unrealistic expectations of the state's growing coal and iron industry. It also ran into troubles arising from Henry's stubborn managerial style, but the Pennsylvania survey eventually served the Pennsylvania coal trade by opening new coalfields and disseminating needed information concerning the state's mineral reserves. The Virginia survey, headed by William Barton Rogers, internalized the growing animosity in the Old Dominion between its western and eastern sections. As had happened with internal improvements, Virginia's political structure colored its geological survey. Both projects ended in the early 1840s due to financial difficulties in their respective states, but in the years in which they operated concurrently the antebellum surveys reinforced the divergent paths of the political economy of coal in Pennsylvania and Virginia. The politics of sectional strife were so ingrained in the Old Dominion that even the state's geology could not escape its grasp. The birds might have "sung coal" in Pottsville, Pennsylvania; in Virginia they sang a different tune.

PATRIOTIC GEOLOGY: THE ORIGINS OF THE PENNSYLVANIA SURVEY

Pennsylvania's first geological survey owed its origins to the same men and institutions that had boosted anthracite coal in the 1820s and 1830s. The early promoter of a survey in Pennsylvania was a Philadelphia attorney and member of the Franklin Institute, Peter Browne, who argued for the formation of a geological survey in the early 1820s but was frustrated at every turn by indifference, ignorance, and outright hostility to his proposals. In an address to the Franklin Institute in September 1826, Browne refined his pitch to emphasize the commercial benefits of the project as well as its scientific merit. The Franklin Institute formed a committee to study the proposal, which recommended "this scheme with earnestness, not only to *individual* but *Legislative* patronage, as meriting in an eminent degree the encouragement of a Republic, *whose immense resources* depend for their development upon a minute and thorough exploration of her soil and natural productions." Thus infused with both academic and patriotic energy, the campaign to undertake a systematic study of Pennsylvania's geology received the backing of the Franklin Institute in 1826.[5]

Although Pennsylvania was not the first state to undertake such a venture, the growing importance of its coal trade marked the Pennsylvania survey as one of national importance. A legislative committee in favor of the survey added self-sufficiency as another benefit of the project. A survey would help uncover Pennsylvanian supplies of "minerals extensively used in the arts, and heretofore imported, either from foreigners or from our sister states," it argued in 1833. An extensive survey would therefore enhance Pennsylvania's reputation as a state with ample reserves of coal, iron, and other important minerals. "We think it fortunate," Benjamin Silliman's *American Journal of Science* commented, "that the first great effort of this kind should be made in a state distinguished for its extent, and for the variety and richness of its mineral productions, and situated so near the geographical centre of the United States."[6]

The formation of the Geological Society of Pennsylvania in 1832 by several prominent citizens of the commonwealth's business and scientific community provided an important boost to the geological survey's cause. This organization wanted to ascertain the "nature and structure of the rock formations of this State" and thus to promote scientific knowledge. It also hoped to find "the uses to which they can be applied in the arts, and their subserviency to the comforts and conveniences of man."[7] Like the Franklin Institute, the Geological Society of Pennsylvania blended state-level patriotism with scientific promotion and counted among its members many influential Philadelphians. The society attempted to gather support for a geological survey in Pennsylvania by organizing the smaller geological associations, encouraging them to petition the legislature, and sending a constant barrage of promotional material to Harrisburg. The Pennsylvania legislature responded by assigning the various petitions, remonstrances, and bills to committees for further review.[8]

The proposed survey rarely emerged from committee without glowing recommendations, but the scope of the project remained unclear. Reports mentioned the scientific value of the survey and acknowledged its academic component, but, as the survey gathered more legislative support, its aims became less and less intellectual. In 1834 a House committee noted that a survey could uncover "hidden treasures" of iron, coal, and salt in Pennsylvania, would attract "Capitalists and other enterprizing individuals," and that "the influx of wealth into the State would be considerable." The report also mentioned that Massachusetts had embarked upon a recent survey and asked that when other states with lesser resources are expending energy to explore their geology, "shall Pennsylvania remain indifferent to her superior advantages?" As the survey moved from the idealistic atmosphere of Philadelphia's academic salons into Harrisburg's less contemplative environment, its patriotic and economic dimensions assumed greater proportions.[9]

The economic value of the geological survey aided its passage through the legislature during the 1835–36 session. In his last message to the legislature in December 1835, outgoing Governor George Wolf boosted the "incalculable value" that a geological survey would have in "developing the hidden resources of our prolific and opulent commonwealth." A committee formed to respond to the governor's endorsement provided glowing assessment of the survey in terms of the state's economic development. In a report authored by Charles Trego, a Philadelphia representative and member of the Geological Society of Pennsylvania, the committee warned that mineral resources "might remain forever unknown and unproductive" unless an experienced geologist were to "detect evidences of mineral wealth in hundreds of places daily passed over by those who are unacquainted with geological and mineralogical science." As a geologist himself, Trego must have been sympathetic to the survey's potential for advancing scientific knowledge. As a legislator, he also knew that science in and of itself would not get the act passed, and his influential report reflects the remunerative potential of the survey.[10]

Because geology was an infant field of science in the United States, passage still seemed uncertain. After all, only a few American geologists were familiar with the theories of European scientists, and many of them still labored as amateurs concerned largely with natural "curiosities" or evidence of the biblical deluge in geological formations. The Geological Society of Pennsylvania eventually became a widely respected geological organization in the United States, although by the 1830s many of its members only reluctantly acknowledged that field's place among the legitimate natural sciences. Alexander Dallas Bache, the first president of the National Academy of Sciences and an influential member of the scientific community, long considered American geologists to be amateurish quacks and opposed their inclusion in Philadelphia's Franklin Institute. Calculating the economic benefits of this survey helped erode the skepticism of many legislators, but more proof of the survey's legitimacy was required.[11]

For the Pennsylvania survey legitimacy arrived in the form of Henry Darwin Rogers, a rising star in American science. He had served as a lecturer at the Maryland Institute in Baltimore, as a professor of chemistry and natural philosophy at Dickenson College, and in 1833 became the first American elected to be a fellow of the Geological Society of London. Two years later he was appointed state geologist of New Jersey, but Pennsylvania's vast mountains and valleys were more alluring to the ambitious Rogers. Although he was not an active member of the Geological Society of Pennsylvania, Rogers emerged as a likely candidate to head the survey on the strength of his European and American credentials and his good standing with Philadelphia's scientific community. Alexander Dallas Bache, despite his disdain for geologists as a whole, respected Rogers and had recently coauthored an article on the chemical properties of Pennsylvania coal with him. Bache helped secure the Franklin Institute's support for his nomination as state geologist. By the time Rogers arrived in Harrisburg to deliver a public lecture on the proposed survey, in January 1836, he was the only serious candidate for the position of state geologist.[12]

With a famous geologist waiting in the wings, the survey's supporters pushed for a three-pronged program. Browne estimated that a comprehensive survey of Pennsylvania, which he divided into twenty-six districts, could be completed in five years at a cost of $3,000 a year. The maps, notes, and reports on each district would be published a year after the survey's end at a cost of $20,000. After the initial survey and published reports, Rogers wanted to create a geological "cabinet" (as collections of mineral specimens for public display were called in the nineteenth century) in each county. An extensive centralized collection would cost upward of $5,000. With the support of Governor Joseph Ritner, iron maker Thaddeus Stevens in the House of Representatives, and internal improvements enthusiast Frederick Fraley in the Senate, the bill establishing the Geological Survey of Pennsylvania appeared in the session of 1835–36. The proposed survey's blend of patriotism and economic utility won over enough skeptics to pass the bill into law in late March 1836. At a time when partisan emotions ran high, the survey slipped through the legislature relatively untainted by party prejudices, demands, or influences. The representatives of western landowners, participants in the anthracite trade, and profit-minded capitalists all saw practical advantages to the survey. In the booming economic climate of 1836 they could afford to be optimistic.[13]

RESTORING GLORY: THE ORIGINS
OF THE VIRGINIA SURVEY

Whereas Pennsylvania's survey began with high expectations, in Virginia the initial impetus to begin a state geological survey took a less sanguine form. By 1830 out-migration from Virginia had drained the state's population, which worried pundits in the Old Dominion. For example, Henry Ruffner, a professor at Washington College in Lexington, estimated that between 1790 and 1840 Virginia had lost more people to migration than had all of the states north of the Mason-Dixon Line put together. Other contemporary observers also noted that young white males, whose most productive years were likely still ahead of them, led the migration. Politicians offered a number of solutions to stave off the loss of population and wealth. Internal improvements, less taxes, better education, and greater political freedom were among the panaceas. A small but vocal antislavery movement advanced emancipation and colonization as a way to convince Virginians to stay put. Their actions led to an open discussion of abolition in the legislature in January 1832 which actually came close to leading Virginia down the road of gradual emancipation. A state geological survey was hardly as controversial as emancipation or as popular as internal improvements, but in the 1830s exploring the state's mineral resources was among the schemes proposed to stop Virginians from leaving the state and restore the Old Dominion to its proper status in the Union.[14]

In 1833 Peter Browne, the tireless supporter of Pennsylvania's survey, returned from a geological excursion in Virginia. Browne had traveled along the state's picturesque Shenandoah Valley to Staunton, crossed the mountains at Jenning's Gap to visit the mineral springs tucked away in the Alleghenies, and then returned eastward, moving through Charlottesville and Richmond on his way to the mouth of the James River. This journey inspired him to write a letter to Governor John Floyd in which he described Virginia's natural wonders, "inviting the mind of man to reflection and his hands to industry," and suggested that the governor initiate a geological survey of his state. Browne emphasized two major elements of Virginia's geology which deserved closer examination. He cited the valuable minerals such as iron ore and bituminous coal which seemed to be present all over the state as well as the curious but popular mineral springs of the western mountains. Browne then pitched the practical benefits of the survey along with its scientific value. This study would not only aid science and increase land values, Browne argued; it would also give "a very proper check to unnatural migrations to the extreme west, by bringing to light and usefulness innumerable valuable crude materials," which would encourage the growth of industrial arts and aid the state's growing internal improvement network.[15]

Governor Floyd gave Browne's suggestion serious attention, as he did most plans to draw his state out of its current dilemma. "It is commerce we need," he told the Virginia legislature in December 1833. "Commerce will enrich any country, though the soil be indifferent; but, when both fertility of soil and a brisk commerce is had, by timely and judicious regulations and improvements, the prosperity of our citizens is secured beyond interruption." Floyd lamented the fact that western counties lay undeveloped and encouraged the legislature to extend roads and canals to the west in order to spur the state's economy. But he also argued that in addition to roads and canals "we ought not to be unmindful of the great wealth which lies buried in the earth, which only requires the examination of men of science to bring before the country, and make known its value and usefulness to capitalists, who would be induced to engage in fitting it for commerce, thereby creating new sources of wealth." In addition to Virginia's iron, gold, and lead reserves, Floyd cited "inexhaustible mines of bituminous coal" as a great potential asset.

The governor then submitted Browne's letter to the consideration of the legislature and left the matter for its members to decide. Later that year Richmond's *Southern Literary Messenger* reprinted the letter and endorsed the idea of a geological survey. "Can the paltry consideration of a few thousand dollars expense," the *Messenger* asked, "outweigh the significant advantages which are likely to result?"[16]

Early in the following session of 1834–35 a number of petitions in favor of the survey convinced the legislature to appoint a committee to study the matter. To determine the desirability of a survey and what shape it might take if funded, the committee drew upon the expertise of the state's most prominent scientists. Among them was William Barton Rogers, the older brother of Henry Darwin Rogers. At the time William was a professor of natural philosophy at the College of William and Mary in Williamsburg. Although he had no formal training in geology, his brother's experiences in abroad with the Geological Society of London filled him with an enthusiasm for the young science. Rogers also had written an article in 1834 in the *Farmers' Register* on deposits of mineral fertilizer in the state. "An active and diligent search," he suggested, "ought forthwith to be commenced throughout the region in which there is a probability that it exists." Like many scientists of the period, Rogers was excited by the prospect of the public employment of geologists. In November 1834 he wrote to his brother Henry that "my reputation in Virginia is rapidly rising, while at the same time a much wider field of exertion seems likely to be open to me." "To obtain such a situation from the legislature, or indeed, to urge them to any measure of the kind," he cautiously added, "will require great activity, not only of me but of all my friends."[17]

Fortunately, Virginia's proposed survey had some influential backers. One of William's close friends was Joseph C. Cabell, a powerful state senator with many political and social contacts and the organizer of the James River and Kanawha Company. Rogers worked closely with Cabell to promote the survey and drew up a glowing report for the legislature's perusal in early 1835. Like his brother, William Barton Rogers was an academic scientist, not a political economist. His report to the legislature, however, mainly discussed the economic benefits of a survey. A geological survey, he argued, would create "discoveries of great general interest and of invaluable local importance" and prove that "the districts of the state, at present almost deserted from supposed meagreness of resources," would reveal "the valuable rocks and ores from which enterprise was to derive new incentives to exertion." In addition to this emphasis on the survey's general economic utility, the report highlighted Virginia's deposits of mineral fertilizer. Whereas iron, coal, and other minerals might spark the interest of some, the vision of revitalizing Virginia's soil with products mined within its borders won strong support in a legislature dominated by agrarian interests.[18]

With the report favorably received, Rogers received a chance to promote the survey by addressing the House of Delegates in person on 9 February 1835. He admitted later to Henry that he had been extremely nervous. Yet, he recalled, he "stood forth boldly and advocated, I think powerfully, the cause of geology, developing a few of its more important truths, and displaying the benefits which it proffered to Virginia." On the strength of Rogers's economic arguments in favor of the survey and his speech to the House, the legislature passed a bill authorizing the Virginia Board of Public Works to appoint someone to undertake a geological reconnaissance of the state and authorized funds for a chemical analysis of the soils, minerals, and mineral waters gathered on that expedition. The bill also restricted spending by the Board of Public Works to fifteen hundred dollars on the reconnaissance and required the geologist to report to the legislature during the next session a "plan for the prosecution of a geological survey of the state."[19]

A reconnaissance of the state offered the state geologist the opportunity to muster the evidence necessary to convince the legislature of the need for a permanent survey. Immediately following the passage of the bill, Rogers was appointed state geologist, and in 1835 his exploration of Virginia began in Newport News, went through Richmond and Charlottesville, entered the mountains at Staunton, and ended at the Ohio River in Huntington. William's report to the legislature of January 1836 summarized his observations and outlined the purpose of Virginia's geological survey. His plan divided the state into five major regions: the tidewater, piedmont, and Shenandoah Valley each constituted geological units, and the Allegheny Mountains were divided into two areas for examination. Rogers's depiction of these five regions highlighted two major economic aims of the survey. First, he spent a great deal of time discussing the state's valuable deposits of marl, green sand, and gypsum—minerals used in restoring the fertility of the soil. These deposits, he argued, "would be an efficacious restorative to the exhausted and sterile soils to which ameliorating applications have of necessity hitherto been denied." In addition to agrarian wealth, Rogers emphasized the industrial potential of Virginia's geology, especially in the western counties. Indeed, he appeared captivated by the economic possibilities of western Virginia. "How magnificent is the picture of the resources of this region," he wrote, "and how exhilarating the contemplation of all the happy influences upon the enterprise, wealth, and intellectual improvement of its inhabitants, which are rapidly to follow the successive development of its inexhaustible mineral possessions."[20]

Virginia's survey, like Pennsylvania's, assumed patriotic dimensions in order to secure legislative approval. William's report highlighted the blend of agricultural and industrial aims of the survey, and he consciously muted the purely scientific aspect of the proposed project. This strategy was successful, as the bill to authorize the survey passed without any problems. The General Assembly authorized the state geologist to hire a few assistants (the survey never had more than five at one time), provided for the placement of specimens in both the state cabinet and Virginia colleges and universities, placed a cap of five thousand dollars on annual expenditures, and authorized annual reports to the legislature. But one striking feature of the bill was that it did not specify how long the survey would be conducted, which was unusual for state geological surveys. Year-to-year funding meant that the survey could last indefinitely. In reality, however, it meant that Rogers had to appeal for appropriations at every legislative session. As its execution demonstrated, this funding structure served as more of a constraint than an advantage to the survey.[21]

From its very inception the expectations of Virginia's survey revolved around economic recovery and, more important, the retention of the state's departing population. In the Old Dominion this meant reviving the state's lagging agricultural sector. Virginia was still the nation's leading producer of tobacco in the 1830s, but it was losing both residents and capital to the booming cotton states of the Southwest. If the state's young people saw a prosperous future closer to home, many Virginians argued, this westward migration could be stemmed. As William Rogers found out, geology had its place in this strategy. The *Richmond Courier and Semi-Weekly Compiler* reported favorably upon establishing a permanent survey in January 1836. "We hope the sons of the Commonwealth, yet remaining will soon see, in the alluring prospects about to be unfolded," the editors pleaded, "enough to induce them to cling to their native soil, as all sufficient to afford them everything which constitutes a just reward for enterprise, industry, and fidelity." Like his brother's endeavor, William Barton Rogers's survey labored under enormous expectations. Henry's mission was to highlight resources that most Pennsylvanians assumed were already present. The Virginia survey, however, was expected to play a major role in the rollback of Virginia's economic decline.[22]

"IT CANNOT HARM US IF WE DO NOT FEEL IT": PROBLEMS IN THE FIELD

Scientific and patriotic rhetoric greased the passage of state geological surveys in both Virginia and Pennsylvania. The ambitious plans created in Harrisburg and Richmond, however, underwent a transformation in the field as the Rogers brothers struggled to reconcile their own scientific interests with the expectations of their legislative sponsors. In Pennsylvania Henry Darwin Rogers seemed careful from the start to blend the academic and economic agendas of the survey, but this uneasy marriage eventually fell apart. In a way Pennsylvania's responsive policy-making structure undermined the survey by allowing critics to hamper its progress. William Barton Rogers's experience in Virginia resulted in less confrontation between science and industry but demonstrated that agrarian interests, especially those in eastern slaveholding counties, dominated its survey. In both cases the priorities of the survey shifted within a few years of their creation but in directions that reflected the distinct political economies of Virginia and Pennsylvania during the 1830s and 1840s.

Whereas both state geologists seemed willing to suppress their academic pursuits in the interests of their legislative patrons to pass their respective projects, the Rogers brothers brought their own expectations to the survey. In their personal correspondence William and Henry revealed an unbridled love of academic pursuits. "Oh, how one may revel in the pleasures of true knowledge!" Henry wrote to William in 1829. "Secluded from men, we may mingle in wider and closer fellowship with Man . . . Only in the deepest privacy can we visit the sealed solitudes of Nature." They also harbored a deep disdain for politics. As an observer of Virginia's constitutional convention in 1829, William bitterly complained that the immense talent and energy of the collected delegates had been misapplied. Politicians should be "elevating the moral nature of our citizens," William complained, and "dispensing truth in all its purifying and ennobling influences." These men devoted their time instead to "selfish passion, to the most futile enquiries, to balancing and counterbalancing local interests, and local prejudices, to complex calculations of numerical ratios, to the appeasement of one section for the transference of its political influence to another, [and] to the discovery of an acceptable quid pro quo for the party concerned." Dealing with politicians demanded a great deal of energy and patience. "There are no [legislative] patrons of genius," William wrote to his brother, "ever ready to assist its efforts, ever active in drawing it forth from the haunts of obscurity and want." Prophetically, he also warned his brother: "You must present yourself before them boldly, frequently and impressively, you must almost obtrude yourself upon their notice: by such good means their good opinion must unfailingly secured."[23]

The Roger brothers tolerated the political demands upon their profession but only because it allowed them to gather critical data for a larger purpose.

In fact, Henry and William hoped to form a new theory of mountain formation based upon their observations in the field as state geologists. They spent
the summer of 1834 in the mountains of western Virginia studying geological
formations and regarded the Appalachian chain as an excellent laboratory
for challenging conventional theories concerning mountains. The prevailing
model in the early nineteenth century held that mountains resulted from either contractions in the earth's crust from cooling or from the vertical pressure of molten rock beneath the crust. Henry and William later argued that
the Appalachians demonstrated that mountains could be formed by a combination of lateral forces and a vertical thrust, which toppled the uppermost
strata and sometimes folded them back upon one another. Such observations,
though important in academic circles, were hardly the objectives that legislative supporters of the geological surveys had in mind. Therefore, the Rogers
brothers endeavored to balance their long-term theoretical objectives with the
short-term expectations of their employers.[24]

Henry planned to use the annual reports required by the state for his political and economic audience while holding back major scientific conclusions
for publication in his final report. Being a firm believer in the scientific
method, he loathed the idea of presenting premature findings in order to satisfy the legislature's short time horizon. In April 1836 Henry advised his
brother that "patience is therefore indispensable, for bear in mind that under
no circumstances ought the State to look for detailed work." Henry's final
commentary on Pennsylvania geology could wait until all the work in the field
was completed. "These popular schemes are too apt to be abortions," Henry
noted of surveys that had bowed to political pressures, "and better do little
than go wrong at the start."[25]

Henry's interests initially dovetailed with those of Pennsylvania's coal industry. His original strategy for the survey divided the state into three regions;
a southeastern region, an Appalachian region that covered the mountainous
ranges stretching from central to northeastern Pennsylvania and contained
the anthracite region, and an Allegheny region that included the state's western counties. Rogers chose the Appalachian region as the initial focus of the
survey in order to appease anthracite interests in Pennsylvania and at the same
time to satisfy his scientific curiosity in theories of mountain elevation. His

first report reflected the synthesis of economic and scientific values. It also re-iterated Rogers's disdain for hasty judgments early in the survey. "A further powerful motive for perfecting our observations before we venture to give them currency," he argued, "is afforded by the reflection, that those conclu-sions at which we are most anxious to arrive, are of a kind bearing necessar-ily upon many private interests, and tending to operate widely on the indus-try and capital of the country."[26]

Whereas his first brief report discussed the typology of the geological strata in Pennsylvania's mountains, Rogers's second annual report included a controversial discussion of the coal formations of Schuylkill County. Gover-nor Ritner had announced at the commencement of the legislative session of 1837–38 that the survey's "operations this season embraces all the anthracite coal beds, and many of the rich iron deposits of the state." The report issued in February 1838 highlighted the region between the Susquehanna and the Delaware rivers and categorized all of its limestone, iron ores, sandstone, and slate. Rogers himself devoted a large part of the year exploring the southern anthracite region for coal. He speculated that many of the coal seams in Schuylkill County had been subjected to a vertical upthrust followed by lat-eral pressure, which crushed them during the process of mountain building. The grinding of the seams made initial attempts to excavate anthracite quite easy—the coal literally tumbled into the cart. Rogers, however, warned that the continued mining of these "nearly vertical and over tilted seams" made the future of cheap mining in Schuylkill County "precarious in a high degree." Rogers noted that "intelligent colliers" had long been aware of this fact but that this tendency of north dipping portions of coal strata to be unproductive in the long run had not been heeded and "injudicious enterprizes are, therefore, often undertaken in total disregard of it." Given the many false starts and failed mining ventures in the anthracite fields, this observation should have come as no surprise to Pennsylvania colliers.[27]

This report highlighted the need for more careful exploration in the an-thracite region, but Rogers had cited what many considered to be the Schuyl-kill coal region's strength—the ease of getting to the coal—as a potential weakness. This was an extremely impolitic move on the part of the state geol-ogist. Although local geologists such as Enoch McGinness set to work immediately to disprove Rogers's discovery and Schuylkill County did not suffer major effects from the report, the political fallout hit Rogers immediately. Following Rogers's presentation of his third annual report, Richard Brodhead of Northampton County introduced a resolution to form a committee to inquire into the further necessity of the survey. The severity of the attack only three years into the survey did not bode well for the project's future. Brodhead's 1839 request never gained momentum in the legislature, but the conservative Democrat from Northampton became a solid enemy of Rogers and the Pennsylvania survey.[28]

Western Pennsylvania coal interests found the survey's annual reports to be of greater general utility but lacking in specific benefits to their economic well-being. In his third report Rogers mentioned a number of potential coal deposits in the west, but his reluctance to make hasty observations gave little guidance for capital investment. In the Stony Creek area of Cambria County, Rogers mentioned coal seams but noted that spring thaws caused the entrance to mines to be blocked during the summer. "We are hence, too frequently precluded from ascertaining the thickness, quality, and aspect of the seams," he wrote, "and from procuring those other data, so essential to trace and develope the coal over the circumjacent country." Rogers was more optimistic in describing the rich Pittsburgh seam, which in his words constituted the "most important and extensively accessible seam of coal in our western coal measures." But, again, the state geologist deferred any detailed analysis until his final report. In his next report Rogers concluded that "several seams, hitherto unknown, were developed" in the northern Allegheny mountains of Pennsylvania, "and it is believed no important deposite within the region has been passed over undiscovered." The same report mentioned the valuable iron and coal deposits of Lycoming, Clinton, and Tioga counties in north-central Pennsylvania as well as the coal measures of the western counties of Armstrong, Clarion, Venango, Butler, Beaver, and Mercer. Initially, western bituminous coal interests clearly had more to gain from the geological survey than did their counterparts in the anthracite regions.[29]

Meanwhile, William's survey began well, but his year-to-year funding forced him to remain vigilant in keeping the survey alive in Virginia's legislature. More so than his brother, who tended to be dour and arrogant, William relished the role of being a public advocate of the geological survey in its infancy. The legislature, in turn, responded positively to William's early speeches. His 1837 address won a $3,900 appropriation for the year as well as praise from local newspapers. William had stressed that it was Virginia's duty to "unfold our natural resources" and that "we should arrest the flood of Emigration—and induce our citizens to remain upon the lands which their fathers had tilled, and which were endeared to them by such powerful associations." The *Richmond Whig and Public Advertiser* reported that William's "picture of the exhaustless stores of wealth and comfort within our own limits, aroused in many bosoms the thrill of patriotism and pride." William constantly reminded the legislature that a geological survey was important to the state's future, but pitching the message in such patriotic terms came at a cost. His brother counseled patience in the face of pressures to produce immediate results. "Heed not the impatience you witness; it is the inevitable consequence of our state of society and institutions," Henry wrote. "It cannot harm us if we do not feel it. The New York survey is ruined by attending the popular impatience."[30]

But William could not escape the political pressures on his survey. At the time of the survey eastern Virginians still controlled the legislature. Economic prosperity in Virginia had always been linked to the soil, so it seemed natural that geological schemes designed to halt the Old Dominion's downward spiral would focus upon agriculture and, more specifically, the search for fertilizing minerals. Heavy use and poor crop management had exhausted the soil in many of Virginia's farms by the early nineteenth century, and some farmers experimented with mineral fertilizers such as gypsum and lime to restore their soil's productivity. Success in these endeavors was limited until 1818, when Edmund Ruffin tested a soft, chalky mineral known as marl on his plantation in Prince George County. His research suggested that marl could increase crop yields by 40 percent, and Ruffin promoted the mineral both as a writer and as the editor of the *Farmers' Register*. He saw the use of fertilizers as a way to increase the productivity of plantations in Virginia, and later in life he earned a national reputation as a strong advocate of slavery in the American South. In this regard the planter-turned-agriculturalist created a unique intersection of Virginia science and politics by advocating geology as a way to preserve slavery. "From the work on calcareous manures, which he has lately given to the world," the president of Hampden Sydney College said of Ruffin to the Historical and Philosophical Society of Virginia in 1833, "it is evident that he is not only a *practical operator,* but a *scientific cultivator.*" Ruffin had no direct role in Virginia's geological survey, but the potential of marl to restore the fertility of Virginia's slaveholding plantations' soil most likely influenced its scope and execution.[31]

The search for marl and other fertilizing minerals constrained William's ability to pursue his interest in the western counties of Virginia. William shared his brother's interest in mountain building and repeatedly emphasized the massive reserves of coal locked away in western Virginia in both his committee report and his reconnaissance of the state. His need, however, to secure financial backing on a yearly basis from an agrarian legislature forced the survey to spend a great deal of time pursuing a more practical form of geology. In the same way that coal and iron boosters in Pennsylvania expected Henry to devote the lion's share of his survey to their interests, the Virginia legislature placed economic concerns above all scientific matters. Unlike his counterpart in Pennsylvania, William was expected to deliver geology's agricultural worth. This emphasis seemed natural. After all, Rogers had worked closely with Ruffin while he was writing his article on Virginia's marl deposits for the *Farmers' Register* in 1834.

The emphasis upon fertilizers is apparent in the composition of Rogers's annual reports. Each of the six annual reports devoted a great deal of space to the discussion of marls, limestone, green sand, and other fertilizing minerals. More often than not, coal took a backseat to marl. The survey's first annual report in 1837 began with a description of the marls of Virginia and contained cautious but glowing reviews of their fertilizing potential. In his next report Rogers noted that the survey was severely delayed because of a lack of assistants and funds but mentioned that one group had concluded an extensive study of marls in the tidewater region. In total four out of the six annual reports focused largely upon the marl region. The exploration of Virginia's mineral assets was thus defined by agrarian interests of the state, as William responded to their demands that geology serve their needs. As the later story of

William's survey demonstrates, this emphasis upon fertilizing minerals came at the expense of western coal regions.[32]

While William's project struggled to balance agricultural, scientific, and industrial interests, the Pennsylvania survey endured attacks from a number of sources. The poor interaction between the survey's assistants and Pennsylvania's colliers, for example, generated a great deal of friction. Henry Darwin Rogers was the head of the survey and ultimately assumed responsibility for its execution, but assistants actually completed the vast majority of surveying, chemical analysis, and mapmaking so essential to the project's completion. Throughout the survey's existence some twenty-five individuals worked under Rogers. Because no formal geological training existed at the time, state geological surveys operated as impromptu academies in which assistants learned to identify and investigate strata, take detailed notes, draw maps in the field, and analyze specimens in the laboratory. The Pennsylvania survey was no exception to this trend, as many assistants who worked on the geological survey went on to lead distinguished careers in geology. Yet the survey required political as well as scientific skills. Assistants on the survey came mostly from middle-class, well-educated backgrounds and displayed a strong passion for their work. They were also, as one historian noted, "apt to be extremely patronizing in their attitude toward the towns and townspeople, miners and mine operators, among whom they did their work." Perhaps his assistants did not realize that, in addition to their geological work for Rogers, they also served as ambassadors for the state survey.[33]

The attitude of J. P. Lesley, who began working for the Pennsylvania survey as an assistant at the tender age of twenty, served as an excellent example of this miscommunication. Lesley used the experience of working with Rogers to emerge as one of the nation's leading geologists, and eventually he headed Pennsylvania's second geological survey, from 1874 to 1888. In 1839, however, Lesley was of the opinion that anthracite miners "keep to the truth about as closely as a pig to cleanliness" and had "minds as base and sordid as the nations from which they came are various," and he was "fully and convinced of the deep depravity reigning here." Some of this attitude might be chalked up to the nativism pervasive among middle-class Pennsylvanians of the time. But nearly four decades later Lesley remained convinced that colliers did not appreciate the value of their survey at the time, because "the language of science was then an unknown tongue, and sounded in the ears of the people like the chattering of animals or idiots." It hardly seems surprising that Pennsylvania coal miners gave Lesley and the other assistants a chilly reception during the survey's fieldwork.[34]

In fact, the latter years of the survey found the Pennsylvania legislature growing more and more skeptical of its worth to the economic future of the state. In April 1838 the legislature appended a rider to a $6,000 appropriation bill which required the state geologist to "make such inquiries and examinations into the present methods of mining coal and manufacturing iron as the Governor shall deem expedient and proper" and to report on this issue as soon as possible. This demand reflected the legislature's interest in nurturing the anthracite iron industry. More specifically, it represented the long-standing hopes of many legislators and colliers that iron ore would be found in close proximity to anthracite fields. "The discovery of iron ore in the immediate vicinity of the Coal, and the plan of smelting it with Anthracite, will open a new prospect to all engaged," the Coal Mining Association of Schuylkill County reported in 1835, "and we may with certainty look forward to the day, when this section will realize the expectations of those who have embarked their hopes in it." In the years following 1838 Rogers tried to include information on iron ores, their location, and their chemical composition in his annual reports. Yet the survey failed to uncover a new cache of iron ore adjacent to anthracite fields. J. P. Lesley later referred to 1839 as the "Iron Year of the First Survey" as a result of legislative pressure to find convenient deposits of iron ore. Later in life Lesley noted that "the best of geological surveys cannot discover what does not exist, nor make available for practical use what costs too much to procure." Rogers most likely found the search for iron veins among the anthracite regions an annoying distraction from the survey's greater work. The task, however, fit the legislature's conception of the state geological survey, and Rogers's failure to produce practical results did not aid his cause.[35]

By 1841 the survey required more money from the legislature for fieldwork. But, as a direct result of the survey's haughty attitude and Henry's inability to discover new coal and iron deposits, the political climate was hardly inviting.

The legislature formed a committee that recommended a twelve thousand dollar appropriation to continue the survey's work. Richard Brodhead, the representative from Northampton County who had attacked the survey two years earlier, served on the committee and submitted a scorching minority report. He claimed that the survey was mismanaging its funds and that Rogers had failed to provide a geological cabinet in each county, as originally promised. Brodhead also invoked Pennsylvania's burgeoning public debt and the need for greater fiscal responsibility. "The people of Pennsylvania have heretofore loved independence and disliked debt, and therefore it would be unjust and inexpedient to borrow money to carry on the survey," the minority report argued. "The Legislature has during its present session, abolished some unnecessary offices, and the work of reform, could, with great advantage to the people, be carried still further." The new appropriation of twelve thousand dollars passed but not without revealing significant dissatisfaction within the legislature. One senator who voted against the bill remarked sarcastically that it "makes no provision for phrenology, physiognomy, animal magnitism, and the highly important science of *water-smelling*."[36]

Henry Rogers, perhaps seeing the writing on the wall, suggested that the survey's work was nearly complete in 1842. That year he presented an abbreviated annual report that discussed the fieldwork that had been completed in the past year along with a discussion of what he planned to do in his final summary. The fieldwork of the final year actually consisted largely of examinations of coal regions, although again the Rogers survey failed to provide detailed information. He only briefly mentioned the "intricate and obscure" Broad Top region in Bedford and Huntingdon counties and claimed that the survey had run out of time in its exploration of the Wyoming and Lackawanna anthracite basins. Rogers described his final report as a detailed summary of the data gathered during the six years of the survey, a series of topographical maps, and a number of geological sections, or cutaway views of the strata. In the spirit of his original proposal Rogers emphasized that "the survey of the State having been mainly undertaken for these economical results, much attention has been devoted to this branch of the investigation" and promised that in the final report every observation made in the field "will be specifically recorded, and wherever it is necessary explained by resort to diagrams and drawings."[37]

With only twenty-two hundred dollars appropriated for the final report from the 1841 act, Rogers thought it imperative to stress the practical worth of the document. If he successfully pitched the final report's utility to coal and iron interests, the legislature would authorize additional funds for its publication. He realized that the state was reluctant to spend more money for the survey and emphasized the importance of an overall summary to the survey's success. In regards to the series of geological sections, Rogers threatened, "should they be withheld, or be imperfectly executed, I shall despair of rendering many parts of our perplexingly intricate, though abundant coal fields, intelligible to the public." With the fieldwork of the survey finally at a satisfactory end, Henry thought he could put the negative publicity behind him and focus on his final report. "It is my firm resolve to occupy two years on my final Report if necessary, rather than further impair my health or let my six years of toil in the field tell for nothing," he wrote to William in late 1841. "I shall, therefore, ask nothing this winter of the Legislature in aid of publication, but report myself of having begun my final report."[38]

In Virginia William Barton Rogers successfully maintained a political balance between his own scientific interests and the agrarian search for rich deposits of fertilizing minerals, but he virtually ignored the mining interests of the state. As the sectional debate over internal improvements suggested, the rift between east and west in Virginia betrayed the underlying tension over slavery's future in the Old Dominion. In the geological survey the undue emphasis on marl in Rogers's annual reports suggests that eastern slaveholding concerns received more than their fair share of attention from the survey. After all, hadn't Browne's original suggestion focused upon the rich mountain areas of the west? Didn't the governor mention the mineral resources of the western mountains when he endorsed the survey? In fact, the survey revealed the process by which, like the political struggles over internal improvements, geology became a zero-sum game in which eastern interests could not benefit unless it was at the expense of western interests. A further examination of the survey's execution demonstrates that it was a game stacked against Virginia's western counties.

Whereas the haughty attitude of survey assistants emerged as a thorny issue in the Pennsylvania survey, William Barton Rogers found his assistants unable to complete the vast majority of fieldwork for the survey. In 1840 William wrote that he often "fretted to find things omitted, which a proper knowledge of the true objects and mode of research should have suggested as important," and he concluded that "one week of my own field work was of more importance to me than months of the comparatively blind labor of my assistants." In fact, the assistants on the Virginia survey brought neither the talent nor experience to their job as those serving in Pennsylvania. Some of them had experience working on other surveys in New York or Ohio, but they did not meet the high standards set by the Rogers brothers for competent geologists. Furthermore, a few of his best assistants were lured away by job offers no sooner than William had trained them.[39]

Assistants assigned to the western counties of Virginia, moreover, found that the insufferable conditions of fieldwork compromised their ability to make accurate and detailed observations. The mountainous terrain of the counties in present-day West Virginia made life miserable and work difficult. "The coal rocks will trouble us," wrote Charles B. Hayden, an assistant looking for coal deposits in Berkeley County. "Not so much from the complexity of structure as from the scarcity of exposures at important points—& the ruggedness of the Mtn's connected with the coal bearing rocks, which it will be necessary to examine *on foot*—*temp 96 degrees* at present—which makes it almost intolerable." Hayden went on to describe sleepless nights caused by "the whole family of pestiferous insects" and observed that the "bedbugs are perfect snapping turtles." Caleb Briggs found that even simple survey work, such as locating exact strata, was difficult. He wrote a letter to Rogers in which he described the isolation of the region and his own financial problems. "I am not only getting ragged but what is far worse I have scarcely money enough to get the holes mended," he wrote in October 1839. "I am certainly in rather a 'bad fix' as the people in this country designate certain situations." Briggs continued to describe the hardships of the field in letters to Rogers. "I ought not in truth to have been in the field," he wrote from Lewisport. "One more summer exposure like this would destroy me." Geological fieldwork in any region involves some hardship, and long-suffering assistants were not exclusive to the Virginia survey. These conditions, however, seriously hampered the ability of Rogers's aides to collect accurate data regarding the coal-bearing regions of

Virginia. Thus, when Rogers wrote his annual reports, the material on the coal regions of the west appeared hopeful but reflected the sporadic and incomplete work of his assistants.[40]

Industrial minerals in the west were not the only victims of the Virginia survey's zero-sum nature. The Richmond coal basin also suffered from the political realities of the survey. Rogers did not assign an assistant to examine the bituminous coal regions of eastern Virginia until 1840. The assistant, Samuel Lewis, was also responsible for examining the piedmont region between the James River and the North Carolina border—a tall order for an inexperienced aide. Rogers's report of that year included a chemical analysis of some of the coals from Chesterfield, Powhatan, Goochland, and Henrico counties, but the samples came from specimens collected earlier, some of which were from mines that were no longer open. He promised that recent collections made by Lewis would provide additional data to "illustrate fully the chemical character and economical value of all the coal seams of this portion of the state." Yet these observations never appeared in print, for the Virginia legislature ended the survey the following year.[41]

William was not without his share of problems in the Virginia legislature. In 1838 opposition revolved mainly around the practical use of the annual reports. "I take for granted that some sneers have been cast upon my labours," William wrote, "and the thought of a legislative body employing itself in venting spleen or exercising wit upon a paper of which but a very few of them have any adequate comprehension, really fills me with indignation." He then observed, "I am at the mercy of the ignorant or the illiberal." In another letter to his brother a year later William described an attack upon the survey in the legislature. "An ignoramus who could not put two words correctly together made an attack in which he attempted to paint me as I addressed the House last year," he wrote, "having a handful of stones before me, such as he could pick up anywhere in the roads in his county; some were red and some were white, and some speckled." "I talked such outlandish lingo that he did not understand a word I said," this country legislator argued of William, "and he doubted whether I did myself." On this particular occasion representatives rose to the challenge and defended the survey, but the outbursts clearly demonstrated that ignorance and the identification of geology with "animal magnetism" and "water smelling" was not restricted to the Pennsylvania Senate. More important, they suggest that members of the Virginia legislature shared to some degree the skepticism of their counterparts in Pennsylvania regarding the practical value of geological surveys. For, no matter how deftly William balanced his reports between his interests in the west and his benefactors in the east, he needed to produce results. An advisor in the State's Auditor Office in Richmond counseled Rogers: "As the legislature *impatiently* look for useful and *practical* results from your labours, it may be well to infuse as much of that information into it as will not mar your plan of a final and complete report."[42]

In March 1841 the legislature voted to fund the survey for only one more year. This decision was due mainly to financial constraints caused by the national depression occurring at the time, coupled with the heavy interest payments required on the state's internal improvement program. Rogers devoted his brief 1842 report to the promotion of his final report. "This report, the crowing work of the survey, from which alone a just estimate of its high economical and scientific value can be formed," Rogers wrote, "it shall be my anxious endeavor to present to the legislature at their next session." Like his brother, William Barton Rogers hoped that his final report would dispel the many aspersions cast upon his public figure and serve as his geological masterpiece.[43]

THE STRUGGLE TO PUBLISH

The ways in which both surveys ended must have left a bitter taste in the mouths of the Rogers brothers. In Pennsylvania the publication of the final report was postponed until 1858. Henry's meticulous work habits caused a delay in the preparation of the manuscript; the survey's political enemies made it difficult for supporters of the survey to wrench funds from the Pennsylvania legislature. In 1843 the legislature offered $2,200 to complete both the promised cabinets and the publication of the final report, but Rogers found this an insufficient amount. In 1845 he argued that the state had already spent upwards of $76,000 and that only a small portion of this amount would be necessary to provide a final report. Additional funds, he promised, would allow him to "array before the reader in a clear light and in detail the distribution of the mineral riches of the commonwealth." Two years later a joint committee of the House and Senate agreed with Rogers that the "extent of the immense mineral wealth of Pennsylvania should be known." Nevertheless, the legislature refused to provide Rogers with the money to publish his final report.[44]

The geological survey found occasional mention in the Pennsylvania legislature over the next ten years, with various petitions and pleas emanating from scientific organizations such as the American Association for the Advancement of Science and the Academy of Natural Sciences. Rogers's political enemies, however, ensured that the Pennsylvania survey received no major allocation until 1855. Charles Trego, an early booster of the survey in the legislature and an assistant under Rogers from 1837 to 1840, published his own book, entitled *A Geography of Pennsylvania,* in 1842 and deliberately omitted Rogers's name in his chapter on geology, although it quite obviously drew upon the survey's work. Trego thought that Rogers had inadequately acknowledged his assistants on the survey and that he rarely gave them credit in both his annual reports and his scientific work outside the survey. While serving in the Pennsylvania legislature in 1840s, Trego worked tirelessly to get back at Rogers by delaying the report. Rogers's abrasive personality and the political enemies it engendered thus hampered the publication of his final report. "A wave of suspicion and dislike, pushed before it by the First Geological Survey through its whole progress," J. P. Lesley recalled, "brought it at last to a dead stop."[45]

Relief for the survey came from the mining industry. As much as colliers resented the way in which Rogers had slighted the Schuylkill anthracite region in his second report, they needed the kind of geological information Rogers had promised in his final report. The state survey's annual reports were among the few comprehensive examinations of Pennsylvania's coal regions, and mining and manufacturing interests wanted a final report. William Parker Foulke, a prominent Philadelphian interested in promoting both science and the coal industry, began a campaign in the legislature to revive the survey. Foulke thought it best to keep Henry Rogers far from Harrisburg and instead hired a former legislator, Charles Hegins, as a lobbyist. The resulting campaign focused almost exclusively upon the developmental aspects of the final report and revived the patriotic rhetoric that had typified the survey's original passage.

Foulke wrote to George S. Hart, a member of the House Ways and Means Committee, that "there is a necessity for availing ourselves of all that science can do in culling out our resources and directing our industry." No more important method of doing so, Foulke argued, is "as concise and effectual for the presentation of our eminent advantages to the attention of the world, as a description of our soils and mineral veins & beds in the form of a geological report."[46]

Meanwhile, Henry offered assistance from Boston, where he had relocated in 1845 to hold a lectureship at Harvard University and to escape his political enemies. Rogers still carried a grudge, and his letters reveal the kind of contempt for Pennsylvania's coal mining industry which Foulke wisely wanted to keep away from Harrisburg. "The miscarriage of almost innumerable enterprizing efforts at Coal Mining and the manufacture of iron," Rogers wrote in 1851, "are equally to be ascribed to the absence of a distinctly written key to the mineral deposits and strata."[47]

With Henry Rogers at a safe distance and the economic value of the survey at the center of the debate, a bill passed in 1851 authorizing publication of the final report. Any doubts about the main sponsors of the legislation were put to rest by the fact that fieldwork was revived in the anthracite region that summer. The joint committee appointed to supervise the publication, however, was a disaster. Political squabbles over who would get the publication contract resulted in a $4,000 loss to the state and the failure to publish anything. By 1854 the *Miners' Journal* reported that $76,760 was spent on the original survey from 1836 to 1842 and that another $32,000 had been allocated in 1851 for a brief renewal of fieldwork and publication expenses. The editors called for a thousand copies of the final report to be made public and warned, "It is high time something should be heard from the one for which such handsome appropriations were made." By 1854 Henry Rogers himself offered to assume the responsibilities of publishing the report. In return for posting bond and providing the state with a thousand copies upon completion, he received the contract in May 1855. That summer he traveled to Scotland to take up a lectureship at the University of Glasgow and to make his revisions to the final report.[48]

In 1858 Henry finally completed his massive tome. *The Geology of Pennsylvania* was issued as a two-volume set, with more than sixteen hundred pages of text in addition to a set of maps and geological sections. The first volume included a description of Pennsylvania's geological regions, the state's various strata and formations, and a detailed account of mountain ranges, minerals, and scenery. Rogers followed through on his promise to note the economic value of Pennsylvania's geology in the second volume of *The Geology of Pennsylvania*. "The truly gigantic scale upon which the coal-fields of Pennsylvania are modeled," he wrote, "gives to this portion of her geology peculiar claims to the attention, not merely of the geologist, but of the capitalist and statesman." Rogers skillfully blended this harmony of interests into his account of both the anthracite and bituminous fields, which focused almost exclusively upon the undeveloped coal strata in Pennsylvania. About the southern anthracite region he claimed that the most valuable deposits were yet unexploited and argued that "a clear exhibition of these will prove of far greater consequence to the present and future industrial welfare of the district than any account of the existing mines or underground workings, however much of temporary interest this information may possess."[49]

Two decades had passed since his slight on the Schuylkill region, and Henry Rogers avoided animosity by focusing upon future potential rather than past mistakes. An earlier show of diplomacy might have led to an earlier final report. Rogers gave explicit instructions for future developers in the western coal regions of Pennsylvania in *The Geology of Pennsylvania*. His discussion of the Indiana County region noted that the "facilities for mining the iron ore, coal, and limestone, in this neighborhood, are certainly considerable; and these minerals all lie convenient to ample water-power." For good measure Rogers appended a "Map of the Anthracite and Bituminous Coalfields of Pennsylvania Exhibiting Their Relation to Various Markets." Positive reviews of *The Geology of Pennsylvania* remarked upon the volumes' optimistic tone and praised the work as a useful guide for capital investment. In providing a practical blueprint for developing the commonwealth's coalfields, *The Geology of Pennsylvania* was highly successful, if not timely. In fact, Rogers's work was reissued in 1868 in response to renewed interest in locating new coal and iron deposits. "No other geological report upon territory previously explored by him can be expected to be altogether original," one reviewer wrote, "or in anywise an improvement upon his fulness and accuracy of detail."[50]

William Barton Rogers's attempt to provide a final report for his survey of Virginia ran into similar problems. Like Pennsylvania, Virginia labored under a heavy debt for its ambitious internal improvements program by the 1840s. After the legislature's last small appropriation in 1842, William turned his attention to his chair in natural philosophy at the University of Virginia and only intermittently tried to convince the legislature to revive the survey. James Brown Jr., an old friend of the survey in the State Auditor's Office, urged in 1844 that William visit the capitol again. "Personal intercourse with the members would be advantageous to the public interest as well as to yourself," he wrote. "You must steal a few days from your darling university at as early a day as possible." Such attempts were in vain, however, as Brown noted that in 1845 only slight opposition defeated the measure to fund a final report. "If you could have spared a few days more from your *professorial* duties I feel confident it would have been successfully carried through," he wrote to William. "The public at large if they but knew it, are deeply interested in having the result of your labours laid before them without unnecessary delay."[51]

William had his hands full with other matters at the University of Virginia. The 1830s and 1840s witnessed major student riots in Charlottesville, and William became deeply involved in their suppression. Violent behavior by the student body was no laughing matter; in the spring of 1839 a professor was horsewhipped in front of his students, and the following semester a student murdered Professor John A. G. Davis, the chairman of the faculty, in cold blood. By 1845 William himself was elected chairman of the faculty—a position that demanded a great deal of time, energy, and tact. Efforts to revive the survey suffered as a result. In 1849 Brown chided Rogers that "there are perhaps not half a dozen persons in the present legislature that are acquainted with the nature of the survey, or that one was even made—The great body of them of course require to be enlightened on the subject."[52]

The lack of a strong advocate, such as the Pennsylvania coal or iron industry, also hampered the push for a final report. Coal interests in Pennsylvania, though they detested Henry Rogers at times, knew that they could use his geological data for their own purposes. Thus, Foulkes and his friends made Henry's work politically palatable in order to secure the publication of the final report. The Virginia survey had no such support. As late as the 1850s, the role of the survey, or geology as a whole, was still very much open to interpretation in Virginia. The *Richmond Whig* published an editorial lamenting the fact that the survey's final report had not been published and noted that few people were even aware that the project existed at all. "Many suppose that geology is merely a search for mineral treasure; or, in other words, that it is an organized hunt for gold and silver, or at least, for mineral wealth of some kind," the editors of the *Whig* argued. "Well, it does embrace these, so far as to ascertain what kinds of minerals do and what do not exist in the State, and to what extent, if at all." But, lest Virginians think that exploiting mineral resources should be the main thrust of the survey, the *Whig* claimed that "the great, the primary object is, or should be, *the better development of the State's Agricultural resources.*"[53]

Promoters of the Pennsylvania survey, while a source of constant annoyance to Henry Rogers, were equally uncertain about their survey's function. Their vision did not earn the respect of scientific geologists such as Henry and William Rogers, but it at least won them a final report. The dream of establishing a technical school and the responsibilities of teaching led William to relocate to Boston in 1853, where he was briefly reunited with Henry. He made a last ditch effort in 1854 to secure funding for his project. The William Barton Rogers who had charmed legislators twenty years earlier was, however, a distant memory. "I am utterly tired of waiting upon the movements of the legislature," he wrote to Henry. "The lobby working, of which I see a good deal and hear more, is as repugnant to my taste as to my sense of right, and I avoid even the colour of it." The bill to provide funds for a final report passed the Senate but could not make it through the House. William left Richmond in March 1854 and promptly embarked upon an extraordinary career in Boston. In 1865 he helped found the Massachusetts Institute of Technology and served as the school's first president. He never completely abandoned his work, however, on Virginia's geology. While giving the commencement address at MIT in 1882, William dropped dead at the podium. His last words reportedly were "bituminous coal."[54]

In Pennsylvania the geological survey coalesced around the uneasy partnership of science, capital, and the state government. Although it labored under heavy expectations and suffered from the leadership style of Henry Darwin Rogers, the Pennsylvania survey must be considered a success in terms of its impact on the state's coal industry. Anthracite miners resented the haughty attitudes of the survey's assistants and its director, but eventually even the most developed areas benefited from the information published in *The Geology of Pennsylvania,* though not without a substantial delay. "Subsequent losses of time, money, and energy in Schuylkill," historian Clifton Yearley writes of the tense relationship between Rogers and Pennsylvania colliers, "were to prove a grim memorial to the breakdown of communication between those who know and those who needed to know." Despite the initial problems with the project, Henry ultimately served the trade well.[55]

The bituminous coal regions undoubtedly profited from the survey. As the least developed and, outside of Pittsburgh, most overlooked coal-producing areas, fields such as the north-central bituminous field in Tioga, Bradford, and Lycoming counties; the Broad Top field in south-central Pennsylvania, and the main bituminous field in the western counties received more attention following the survey's completion. In those areas the state geological survey truly served the purpose of underwriting costs for mineral development, as the information gleaned from Rogers's annual reports and *The Geology of Pennsylvania* aided colliers throughout the nineteenth century. In the end the survey's findings made western Pennsylvania appear less remote and less mysterious in terms of its coal reserves, and, for capitalists, a less risky venture.[56]

The Virginia survey's legacy stands in direct contrast. Although William Barton Rogers's role as state geologist and professor at the University of Virginia ushered in a new era of science in the Old Dominion, the geological survey failed to restore Virginia to its old position of economic leadership. Mineral fertilizers had only a moderate impact upon Virginia plantation agriculture. Edmund Ruffin and various agricultural reformers continued to promote marl and other mineral fertilizers, but they did so without the aid of the state geologist. The Richmond basin continued to suffer without the benefits of a professional survey. Unlike Pennsylvania's anthracite colliers, who eventually incorporated the survey's geological knowledge into their industry, Richmond area miners continued to use speculative mining practices through the late nineteenth century. The most telling failure of the Virginia survey occurred in the trans-Allegheny region. Sectional politics torpedoed William's interest in that part of Virginia, and many areas in the west remained unexplored for years to come. Just as the west suffered in internal improvements, the zero-sum game of the Virginia antebellum politics obscured the immediate value and future promise of its mineral resources by skewing the geological survey in favor of its eastern constituents.

The state geological survey was a peculiar product of its times: wide-reaching in scope but personal in structure. In the early years of the projects William's charisma and Henry's impeccable credentials aided the cause of the state surveys immeasurably. Over time William's weariness of Richmond's endless bluster and political posturing killed Virginia's survey, while Henry's lack of tact nearly destroyed the effort to make his research accessible to the public. It is a testament to the profound difference in the policy regimes of each state that Pennsylvania's more flexible system accommodated the dour and difficult Henry, whereas Virginia's political structure doomed the more diplomatic William to fail in the quest for a final report.

In many ways these scientific projects in Virginia and Pennsylvania resembled internal improvement systems in each state and their interaction with the coal trade. In Pennsylvania the legislature launched the State Works with patriotic zeal and expectations of immediate economic gain. As that system, like the survey, failed to live up to its lofty expectations, it weathered accusations of fraud, corruption, and uselessness from myriad critics. In the end, however, the state's jumbled web of public and private improvements allowed Pennsylvania's coal trade to expand. The same could be said for Henry Darwin Rogers's experience with the survey. Throughout the antebellum era Harrisburg specialized in ungraceful, but efficacious, policies regarding mineral development.

The story of Virginia survey offers a striking parallel with that of the James River and Kanawha Canal. The JRKC supposedly linked the disparate regions of the Old Dominion and served as a unifying centerpiece of Virginia's internal improvements. In reality sectional problems ravaged the JRKC, and the project emerged not as a symbol of Virginia's unity but, instead, as a favored child of the eastern slaveholding plantations. The complaints that western Virginians voiced about the JRKC project applied to those of the geological survey as well: there was less money, little interest, and no intention of executing the program for the benefit of western industrial interests. Once again the West would have to fend for itself without Richmond's patronage, and once again the sectional debate in the legislature served as the divisive issue.

The state geological survey, along with internal improvements, constitutes a second way in which the intersection between private interests and public policy shaped the development of mineral resources in Virginia and Pennsylvania. By reflecting the legislative politics of Richmond and Harrisburg, the surveys played a major role in the divergent paths of each state's political economy of coal. The institutional characteristics of each state's transportation networks and geological surveys also emerged in the corporate chartering policy of Virginia and Pennsylvania. As the next chapter demonstrates, Pennsylvania's muddled competence and Virginia's sectional divisiveness shaped not merely the infrastructure or information required by coal mining firms but also the very structure of the firms themselves.

Miners WITHOUT Souls

Corporations and Coal in Pennsylvania and Virginia

*I*n January 1837 Senator William Smith of Culpepper County, an ardent Jacksonian Democrat, led a spirited opposition in the Virginia Senate to a bill designed to streamline the process of chartering mining and manu-facturing companies in the Old Dominion. Although he was a champion of states' rights, Smith acquired the nickname "Extra Billy" from the frequent bonuses he received from the United States Postal Service as the owner of a mail coach business. Despite his penchant for working with federal institu-tions, in Virginia Smith constantly "waged a bold and determined war against the banks of the State, and denounced the whole banking system as inefficient and inherently corrupt," according to his memoirs. In response to the 1837 bill he repeatedly introduced clauses designed to limit the life of charters and tried to shelve the legislation indefinitely.[1]

Smith and his anti-charter colleagues viewed the rapid growth of corporations as an unsettling development. During the same session an anti-charter writer in the *Richmond Enquirer* warned that a General Assembly dominated by corporate interests would create "a splendid country checkered with rail-roads and canals, here and there ornamented with a palace, but with a poor, oppressed *community* of people, governed by combinations of privi-leged corporations, and pampered stock-holders and stock-jobbers." Al-though the political career of Extra Billy soared in the later years to include two stints as Virginia's governor, he lost on this particular measure, as the cor-porate chartering bill passed into law on 13 February 1837. But, despite the suc-cess of this bill, anti-charter ideologues such as Extra Billy remained active in Richmond throughout the antebellum period.[2]

A year later the Pennsylvania legislature struggled with a contentious char-tering issue of similar stature. Representatives of the Offerman Mining Com-pany of the southern anthracite district asked the legislature for a corporate charter that combined the right to mine coal and the right to build their own railroad. This request might seem innocuous, given the wide range of privi-leges afforded to many nineteenth-century corporations, but the coal inter-ests of Schuylkill County fought the measure tooth and nail. They sent peti-tions to Harrisburg, led raucous anticorporation meetings in which laborers asked for protection against "the *rapacity of bodies without souls*," and im-plored their representatives to kill the charter. Pennsylvania's Anti-Masonic Governor Joseph Ritner, a self-proclaimed foe of special privilege, vetoed the bill despite its passage in both houses. When the Pennsylvania Senate over-rode the governor's veto, Schuylkill County's *Miners' Journal* announced that "our representatives have deserted the county and gone over to be the repre-sentative of a few stock jobbers and speculators, some living in, and some out of the state. . . . Like him who betrayed his Master for a few pieces of silver, they will richly deserve the fate which awaits them." When nine more mining firms were chartered later in the legislative session, the editors exclaimed, "Was there ever before such an instance of public prostitution!"[3]

These political struggles over the use of corporate charters by mining firms occurred at a critical juncture in the development of the U.S. coal trade. Whereas early miners successfully raised coal through crude and labor-intensive methods such as trenching or drift mining, small-scale production

made few mine operators rich and placed serious constraints on future production. As entrepreneurs realized that profitable coal mining entailed large
start-up costs, for digging shafts, pumping water, and constructing links with
major thoroughfares, they looked for ways in which to increase capital investment and minimize risk. As a result, in Virginia and Pennsylvania coal
mining became one of the first industrial pursuits to use the corporate form
of organization. But the use of the corporate form did not occur without some
controversy. The political nature of corporate chartering, as the opponents of
Extra Billy Smith and the representatives of the Offerman Mining Company
could attest, challenged the rapid adoption of corporate forms of organization in each state.[4]

Colliers in Pennsylvania and Virginia attempted to build some semblance
of order amid the political chaos that often surrounded the creation of corporations. Whether they fought the generous conditions of a rivals' charter or
pressed for favorable terms of incorporation themselves, coal mining interests
in each state became enmeshed in the chartering policies of their state. As with
the other elements of Pennsylvania's political economy of coal, chartering policy took the controversial and seemingly irrational turns endemic to a more
open and geographically distributed legislature. Virginia's chartering system
followed a more conservative course. Just as sectional disputes disrupted the
development of internal improvements and agricultural concerns dominated
its geological survey, the reliance upon local institutions restricted the effec-
tiveness of its chartering regime. Both states dealt with the growth of the industrial corporation by using the institutional tools available at the time, and
the course of chartering policy prodded Pennsylvania and Virginia's colliers
in increasingly different directions. A localized chartering policy that grew out
of Richmond's politically conservative atmosphere absolved legislators in Virginia of the time and trouble involved in creating corporations. But, as Pennsylvania's more frenzied course of legislative action demonstrated, an active
hand in chartering could allow the coal trade to expand to unprecedented levels. Institutional frameworks in each state shaped chartering according to
their own distinct contours; the end result was greater divergence between
Pennsylvania's and Virginia's political economy of coal.

THE NINETEENTH-CENTURY CORPORATION

It is difficult to imagine U.S. industrial development without the corporation. The large-scale capitalization, limited liability, and minimized risk of the corporation revolutionized business in the United States over the course of the nineteenth century. But a benign attitude toward the corporate charter is a relatively recent phenomenon. Intellectuals, legislatures, and entrepreneurs all struggled to reconcile the vast advantages afforded by corporations with the republican rhetoric of antebellum America. In 1820 the political economist Daniel Raymond denounced the "artificial power" that corporations bestowed upon its shareholders and ruled that "all money corporations are detrimental to national wealth." Rampant chartering at the state level discounted this anticorporate conventional wisdom at the same time that Jacksonian Democrats waged war against the "Monster Bank." But the corporate charter also had their champions during this time. Massachusetts Senator Abbott Lawrence ridiculed those who doubted the "millions and millions of dollars [that] have been added to the State's added wealth" by manufacturing corporations. "If you can longer doubt the expediency of encouraging corporations," he argued in 1836, "I should think you incurably blind, and sealed against conviction." Across the nation legislators dipped from a curious stew of ideological principles, economic interests, and partisan rhetoric when they considered the various bills created, extending, or killing corporations.[5]

To add to this political confusion, a balkanized system of chartering generated a great deal of regional and even state-by-state variation in the creation of corporations. Nineteenth-century jurists and legislators used the rather forbidding term *foreign corporations* to describe firms chartered in other states and often upheld or passed laws restricting their operations. Court decisions preserving corporate rights gradually whittled the hostility embedded in the term *foreign* into a more formal distinction over the course of the nineteenth century, and by the 1880s the landmark chartering laws of New Jersey signaled the rise of the truly national corporate charter. Since then, differences between the corporate charters of each state have declined.[6]

For most of the nineteenth century, however, each state forged its own corporate chartering policy by blending traditional republican ideology, practi-

cal concerns over economic development, and the desire to compete in interstate markets for goods and capital. Leadership among important states such
as New York, Massachusetts, and Pennsylvania became evident by the 1850s,
but important regional and even state-by-state diversity persisted. The voting
rights of stockholders varied greatly among states, for example, with important ramifications for the internal governance of corporate firms. The point
at which each state allowed for "general" incorporation, in which charters did
not require a separate act of the legislature, also varied. As early as 1811, New
York had a general manufacturing law, but general incorporation did not become mandatory for Pennsylvania until 1873. In other words, a separate act of
the legislature was required to create each corporate charter in many states.
The charter had to be introduced and voted upon to become a public law.
Louisiana was the first state to abolish this method of granting charters in 1845,
and by 1850 only six other states had done the same.[7]

Throughout the antebellum period, then, special acts remained the most
common method of granting corporate charters. Because private corporations evoked images of privileged profit at the expense of society, the majority of corporate charters issued in the early United States went to companies
that built turnpikes, bridge companies, and canals. Internal improvement
projects were easy to justify in that they provided a public good—theoreti-
cally, all citizens benefited from the construction of a road or a canal. Other
enterprises requiring large-scale capital, such as banks and insurance companies, also needed charters, although the compatibility of these institutions of
privilege and power with a democratic society bothered traditional republicans. Legislatures were similarly reluctant to issue charters for mining and
manufacturing concerns, especially since these industries already existed
without corporate privileges. Some early firms in this area of enterprise received charters but to a lesser extent than transportation companies. Pennsylvania granted just nine manufacturing charters to companies from 1814 to
1816, and Virginia did not pass a mining or manufacturing charter until 1828.[8]

Charters emerged as a common feature of the legislative landscape in the
years following the War of 1812, but corporations were steeped in controversy
during the first half-century of American independence as legislators sorted
out the public and private needs involved in the process. The transportation

boom of the 1820s and 1830s ensured that the passage of charters occupied a great deal of any state legislature's time and effort. Whereas charters for bridges, canals, and even railroads fell under the guise of public improvements, a creeping mistrust of the "soulless" character of corporate entities permeated the political culture of the times. The use of lobbyists, logrolling, and outright bribery to pass charters raised even more suspicion. By the time President Andrew Jackson announced he would veto the renewal of the charter of the Second Bank of the United States in 1832, the role of corporations in a democratic society faced serious political challenges. The story of Andrew Jackson's struggle against Nicholas Biddle's "monster" corporation in the halls of Congress is a well-chronicled episode of American history.[9]

But the "Bank War" of the 1830s also raged in the state legislatures, with real policy implications for corporations in general and mining firms specifically. In Pennsylvania Jacksonian Democrats maintained an uneasy truce with the state's banks; they were eager to make political capital from their denunciation but also wary of the economic and financial problems an extreme anti-bank stance would provoke. In Virginia the Bank War created a similar environment. Whigs cautiously supported the chartering of banks, and Democrats broke into radical anti-bank and conservative factions. In both states the Bank War exposed corporate chartering to intense political scrutiny and played a major role in determining the shape of corporate chartering in the coal trade.[10]

EARLY MINING CORPORATIONS IN PENNSYLVANIA AND VIRGINIA

A debate over corporate chartering raged in Pennsylvania's anthracite fields during the Jacksonian period, and one of the earliest political actions against incorporation targeted "foreign" interests operating in the state. During the speculative boom in coal land during the late 1820s and early 1830s, firms chartered in New York purchased and held land in the anthracite regions. Local mining interests argued that these foreign corporations were speculative ventures that held back the development of the anthracite trade by keeping valuable coal land from miners. Because these firms were chartered in another state, they were able to avoid Pennsylvania statutes of mortmain—which prohibited land from being held stagnant for extended periods of time. Years earlier Senator Stephen Duncan of Philadelphia had provided a strict interpretation of chartering authority in response to complaints by mining interests about foreign corporations. "The power of incorporation is a sovereign power," he reported to the legislature in 1825, "and it will not be pretended that the legislature of New York can exercise sovereign power within the territory of Pennsylvania."[11]

A new barrage of complaints from the anthracite regions during the early 1830s convinced the legislature that the future interests of Pennsylvania were at stake. It responded in early 1833 by passing an act that allowed the state to confiscate the land of foreign corporations. Henceforth, Pennsylvania prohibited "any corporation to prevent or impede the circulation of landed property from man to man, without a license from the commonwealth," ruled that companies formed in other states did not have the authority to hold Pennsylvania land in perpetuity, and warned that if these firms did not secure a Pennsylvania charter they would forfeit their lands to the state. The Delaware Coal Company and the North American Coal Company, both of which operated under New York charters and claimed to have most of their stock in the hands of Pennsylvanians, were allowed three years to secure a Pennsylvania charter or sell off their interests. A few firms secured Pennsylvania charters and continued to operate, but the lasting legacy of the 1833 Escheat Act was its strong policy implications regarding foreign corporations operating in Pennsylvania. Reflecting concerns that New York might dominate the commerce and industry of Pennsylvania, the legislature clearly announced that mining corporations operating in Pennsylvania would have to receive their authority from Harrisburg.[12]

Another check on corporate power in mining came as a result of competition between the Schuylkill and Lehigh interests. The charter of the Lehigh Coal and Navigation Company (LCNC) contained the right both to mine and carry coal, whereas the legislature prohibited the Schuylkill Navigation Company from owning or mining coal lands. The different privileges of the two carrying systems formed the crux of a fierce debate over the future role of corporations in the Pennsylvania anthracite fields. At first the Schuylkill coal interests had no particular problem with corporations. After a string of Schuylkill County charters failed in the legislature, however, as a result of both the efforts of Lehigh interests and general anti-charter sentiment, many local boosters in the Schuylkill area began to denounce corporations in general.[13]

The battle between the two regions spilled into the Pennsylvania legislature during the early 1830s. Petitions from Schuylkill County interests accusing the Lehigh Coal and Navigation Company of misuse of their charter began to circulate around Harrisburg. Josiah White, the president of the LCNC, responded to accusations of excessive tolls and other abuses with a printed circular. White tried to put a patriotic spin on the LCNC's corporate status. "Our stock is owned by citizens of several states, but principally in our own state. I believe none by foreigners. Our stock is fifty dollars per share, which is possessed by a great number of people; a considerable portion by the working class—the widow and the orphan," White wrote in February 1832. "I presume 10,000 souls are this moment supported more or less by the outlays of this concern, or injured by the long suspension of dividends."[14] Attacks on chartered interests, in the context of Jackson's anti-bank campaign, sometimes made for good politics in Pennsylvania. White followed his 1832 circular with a more detailed appeal to the Pennsylvania Senate's Committee on Corporations the following year. He again responded to the specific charges of high tolls and land speculation as well as the anticorporate attack upon the LCNC. "Petitions, you know, are easily got up, but more particularly, when got up under the specious cry, of 'down with monopolies,' 'down with aristocracies,' &c," wrote White. Despite these attempts to defuse the controversy, the legislature continued to investigate the LCNC and any alleged charter violations.[15]

Leading the way in the pamphlet campaign against the LCNC and corporate privilege in general was George Taylor, an influential writer in the region and the retired editor of Pottsville's *Miners' Journal*. In 1833 Taylor wrote a pamphlet describing the impact of incorporated companies on Pennsylvania's anthracite coal trade. He argued that not only was the coal trade plagued by mismanaged and greedy joint stock companies but that "such institutions are injurious, inexpedient, and unnecessary" for the future of the trade. Taylor predicted that continued chartering of corporations would "be productive of great injustice, and of ruin to the individual operators engaged in the Coal Trade, now forming a large population," citing Adam Smith as an anticorporate authority. He also was careful to distinguish between corporate capital in the carrying trade, which he found completely legitimate, and its presence in coal mining, which was absolutely not. "It must be held in remembrance that in this contest it is the capitalists who are striving to get exclusive privileges," Taylor explained, "to the injury of the individuals who merely ask to be let alone, and that their rights and property may be protected." Poor management, the combination of land ownership and carrying privileges, and the fact that corporate privileges promoted inequality among producers combined to make charters completely incompatible with anthracite coal mining, in Taylor's mind. In the regions without corporate firms, he argued, an "individual Coal owner moves with his family into the Coal Region to make it his home and permanent place of residence" and replaces the temporary shelters of the corporate mining firms with "permanent enlightened communities" that contributed to "the honor, the strength, and resources of the commonwealth." In addition to the corporation's deleterious effects upon community development, Taylor's characterization of corporate mining painted chartered firms as soulless institutions that tended naturally toward monopoly. His words echoed the anticorporate positions of some of the most influential political economy texts of the day.[16]

Whether the struggle over the future of the corporation in Pennsylvania's anthracite fields was an ideological contest between corporate and individual ownership or a competition between rival business interests for the lion's share of the anthracite carrying trade is uncertain. An influential maxim concerning the use of corporate charters in mining interests did emerge, however, from these political contests. During the controversy over the firms operating in Pennsylvania under a New York charter, for example, Senator Duncan noted that compatibility of corporate charters of any origin and coal mining seemed dubious. Incorporation was fine for canals and turnpike roads, he argued in 1825, but corporate chartering for industries within the reach of proprietary firms "is not only inconsistent with the dictates of sound political economy, but at open war with the principles of a free government, instituted for the purpose of promoting the common welfare by the security of equal laws, equal privileges and equal rights." Duncan agreed with the New York companies that a greater amount of coal might be brought to market if corporations were allowed to develop coalfields. But he argued that corporate development would drive individual entrepreneurs out of coal mining. The "partial and transient benefits" of increased production, Duncan maintained, should not be "sustained at the expense of the permanent welfare of the public."[17]

The clear articulation of an anti-charter policy in coal mining appeared in the Pennsylvania Senate's 1834 investigative report on the anthracite industry. Responding to the war of words between the LCNC and the Schuylkill County interests, the legislature authorized a committee to investigate both the past and present condition of the coal trade and also to "inquire what further legislative provisions are necessary to protect, facilitate, and encourage this branch of industry." Chairing the committee was Senator Samuel J. Packer of Northumberland County. Packer's report began with a descriptive account of the various fields, the individuals and companies working them, and the transportation facilities to urban markets. Packer noted the wide privileges contained in the LCNC charter: "Every charter or act of incorporation, is to a greater or less extent an infringement upon the natural rights and liberties of the people—and their natural tendency is to monopoly." The committee recommended that the state purchase the Lehigh Canal and its coal privileges in order to remove the baneful effects of corporate privilege in mining. In order to justify this action, Packer used patriotic logic. "By judiciously fostering our mineral resources," the committee reported, "we may place our State upon a basis too solid to be shaken."[18]

Financial constraints ruled out purchase of the LCNC by the state, but the Packer committee's report nonetheless established guidelines for future chartering policy in coal mining. "There is at this day no greater necessity for conferring corporate powers upon a class of men to mine coal," they argued, than there was thirty years ago "to enable a society of carpenters to plane boards, or of farmers to plough their lands." The committee drew upon the public service interpretation of corporate endeavors to justify this line of reasoning. Corporations carrying coal to market provided a public service, it maintained, while those that owned the land and monopolized the extraction of coal did not. The aim, then, of the Packer committee was not limited to investigating whether the LCNC and any other corporations were currently exceeding their charter; it also attempted to erect a more open system of exploiting Pennsylvania's mineral resources in the future. To further support their claims, the committee attached the testimony of a number of coal dealers regarding the role of corporations in coal mining.[19]

The Senate committee gave pro-corporate interests an opportunity to defend their actions but presented overwhelming evidence in favor of the anti-corporate position. Representatives of the LCNC argued, for example, that individuals might be able to economize more than a corporation, but the capital required for the business of mining and transporting coal was beyond the reach of most individuals. Besides, the LCNC argued, superior managerial skills were not consistent across individual operators, "while companies may always select their officers for their peculiar qualifications." But the LCNC represented the minority view. Most of the miners and coal dealers from across the state's anthracite regions echoed the committee's opinion that corporations tend toward monopoly and represent a menace to democratic society. "We do not believe putting a district or country under the control of an incorporated company will produce either independence of character or freedom of thought or action in its inhabitants," argued coal dealer Dan Bennett of Schuylkill County. "On the contrary, when they are dependent on one source alone for bread, they will soon become so for their opinions also."[20]

Although the foreign corporations controversy and the legislative investigation of the LCNC occurred as a response to intense competition between anthracite regions, they shaped the thinking of future policy makers. First, any corporate endeavor in the Pennsylvania coal industry was expected to have a charter from Harrisburg. This fact would become more important later, with the advent of general incorporation in Pennsylvania, when New York and Massachusetts capital flooded the anthracite region. Individuals also successfully engaged in coal mining throughout Pennsylvania, so many legislators concurred with the Packer committee that corporate charters were not necessary for mining coal. This animosity did not completely eliminate coal mining charters, but it did create significant barriers to the incorporation of mining companies.

The political machinations needed to create a corporation also discouraged mining firms from seeking a charter. During their journey from bill to law, many charters encountered legislative resistance. Representatives and senators hostile to coal corporations undermined charters by voting against them, attaching untenable amendments, or preventing them from ever escaping committee. Legislators also used "pinch bills," in which they would introduce a rival bill simply to extract a bribe from the potential incorporator. Another common weapon for the anti-charter legislator entailed striking out the charter's limited liability clause or adding an amendment that made stockholders liable for all debts and claims against the company.[21] These difficulties could emerge as a result of local interests or from a politician's general antipathy toward the corporation. Roll calls on chartering votes were rare, so it is difficult for historians to reconstruct which legislators consistently voted against corporations and which ones supported them. One quantitative analysis of state legislatures from 1833 to 1843 found that votes on corporations displayed weak partisanship, as many Whigs and Democrats tended to vote similarly on corporate bills. Another study of Pennsylvania suggested that geographical region was the best predictor of how legislators voted on issues of economic development. Whatever the origin of the animosity, aspiring corporate interests needed both political and financial power to survive the chartering process. The seventeen corporations that received charters in the Lehigh region from 1820 to 1845 maintained a close relationship with the Lehigh Coal and Navigation Company and used the LCNC's political power to secure the passage of their charters.[22]

Even if a bill escaped both the House and Senate, Pennsylvania's governors often vetoed corporate charters during the antebellum era, as Governor Ritner did in the Offerman Mining Company's quest for a charter in 1838. As the executive officer in charge of the interests of the commonwealth as a whole, the governor was able to rise rhetorically above the logrolling and underhanded deals that often accompanied legislative politics. In 1842, for example, Governor David Porter condemned corporations in his annual message to the legislature: "They absolve men from personal liability," he declared, "and may tend, by undue combinations and concentrated action, to embarrass the operation of government and interfere with the popular sovereignty." In 1845 Governor Francis Shunk vetoed a charter for the North Branch Railroad and Coal Company and, citing Schuylkill County as proof, noted that private enterprise was sufficient: "Among the varied pursuits of men there is, perhaps, none more simple, or more completely within the compass of individual resources, than that of mining coal." Although charters sometimes passed over the wishes of the governor, the veto emerged as another significant hurdle for coal corporations in antebellum Pennsylvania.[23]

Public pressure from Schuylkill County interests, often led by the *Miners' Journal,* also hampered chartering during this period. Benjamin Bannan, who assumed control of George Taylor's financially weak *Miners' Journal* in 1829 at the age of twenty-two, renewed the journal's anti-charter enthusiasm. Through the publication of mining statistics, acerbic screeds against the free trade movement, notes on new mining technology, and general commentary on the state of the trade, the *Miners' Journal* became anthracite coal's leading trade journal. Personally, Bannan was a strong believer in protectionism, moral reform, and the sanctity of individual entrepreneurship. Throughout his editorial career he also displayed contempt for the Democratic Party. As editor Bannan ensured that any bill that threatened Schuylkill County's system of individual enterprise was exposed and condemned. The *Miners' Journal*'s influence carried well beyond Schuylkill County—its columns were read and reprinted across the state. Furthermore, Bannan's animosity toward the Democratic Party gave the *Miners' Journal* a rather atypical Whig anticorporate editorial line. With Bannan leading the way, anticorporate rhetoric from Schuylkill County during the 1830s and 1840s proved fierce, unrelenting, and bipartisan.[24]

The ideological pervasiveness of the corporate chartering issue in Pennsylvania was also apparent in the Constitutional Convention of 1837. Amid the general clamor for the reform of gubernatorial power, judicial appointments, the voting franchise, and public education, the role of banks and corporate charters in the state's future served as the centerpiece of debate. When the convention opened in the spring of 1837, delegate Charles Ingersoll led the anti-charter forces. "All corporations, especially of perpetuities, conferring privileges for gain are unrepublican and radically wrong," he argued from the floor of the convention. "Equality is put an end to, and an aristocracy is created, which, although without titles, must be inconsistent with the genius and principles of free institutions." When the convention voted on whether to publish Ingersoll's views as the minority report of the Special Committee on the Currency, Corporations, the Public Highways, and Eminent Domain of the State, the future of corporate chartering in Pennsylvania seemed in doubt. Although conservative delegates denounced Ingersoll's report as that of "raw Irishmen and imported democrats," the convention body only narrowly voted down its publication, sixty-eight to fifty-seven.[25]

The convention's discussion of corporations did not result in any dramatic reversals in the state's chartering policy, but delegates did impose some important restrictions on the chartering process. After 1838 the legislature could not create, renew or extend bank charters without six months' public notice. Another reform limited private corporations to twenty-year charters. Finally, the legislature was given the constitutional right to alter, revoke, or amend any charter that it might grant. The end result of the proceedings suggests that most delegates considered banks and other corporations a necessary evil but also as entities that merited the constant attention of government officials. In the end Ingersoll's forces lost the battle to restrict corporations in Pennsylvania severely, but anti-charter forces succeeded in making the danger of rampant chartering evident to the convention.[26]

The enemies of corporate privileges in Pennsylvania boasted no formal party or movement, but the informal constraints on certain types of charters remained formidable after the Constitutional Convention. In terms of coal mining, anti-charter forces permitted relatively few charters for coal mining companies to pass through the legislature during the Jacksonian era. From 1834 to 1854 only 48 companies (about 3 per year on average) were chartered with the right to mine coal, whereas over this same period the Pennsylvania legislature passed 1,089 business charters.[27]

Although Virginia's General Assembly created transportation, banking, and insurance corporations during the early nineteenth century, the use of corporate charters by manufacturing or mining concerns was uncommon until the late 1820s. Bereft of large urban centers and dominated by plantation agriculture, Virginia's limited industrial economy of the early nineteenth century created little incentive for firms to pursue the limited liability, sizable capitalization, and diversified risk offered by corporate charters. But even the Old Dominion witnessed an increase in chartering activity during the Jacksonian era. When the charter for the Rappahannock Manufacturing Company passed in 1828 after several days' debate, the editors of *Niles' Register* celebrated the "time when the representatives of the state felt willing to let the old aristocracy 'go by the board' and raise up an invaluable class of *productive* persons, to be the glory and defence of the state." During the 1830s the legislature passed more and more charters for industrial firms; from 1835 to 1840 it pushed through 172 mining and manufacturing charters, compared to 46 in the first five years of the decade.[28]

Individual proprietors and small partnerships dominated the coal industry during the Jacksonian period, so coal mining firms were not common recipients of corporate charters. John Heth, the son of Harry Heth, and his partners successfully petitioned the legislature for a charter in 1833 to create the "Black Heath Company of Colliers," which was the first coal mining corporation in Virginia. The following years saw a dramatic increase in the number of charters issued for mining firms but not for the coal trade. The discovery of gold in northeastern Virginia in the early 1830s triggered a flurry of petitions to the legislature for gold mining charters. In Culpepper County alone, thirteen firms were chartered to mine for gold by 1837. Virginia's antebellum gold rush accounted for the lion's share of mining charters during this period, as only four of the twenty-three mining firms chartered in 1834 and 1835 were granted for the purpose of mining coal. Gold miners crushed and rolled the rock surrounding gold veins and then washed the gold out of the debris, which made the excavation of precious metals a capital-intensive industry. Not surprisingly, gold mining operations in the Old Dominion used the corporate form of organization in order to acquire large tracts of land, hire a workforce, and purchase machinery. Here the similarities between gold and coal mining end, for the skill and technology necessary to refine gold has little application in the coal trade. Also, the high value of a small amount of gold meant that cheap transportation to market was not critical to gold mining operations. In terms of shared knowledge or machinery, Virginia's gold mining corporations had little to share with the state's colliers.[29]

But the Old Dominion's gold rush proved short-lived, which may have suggested to many Virginians that mining corporations served speculative, not productive, interests. This interpretation of chartered firms blended well with the traditional misgivings of corporate ventures. A group of proprietary colliers in the Richmond basin complained that Heth's 1833 charter demonstrated "no privilege can be *given* to one or more Citizens that does not to the same degree *take* from the rights & privileges of their fellow Citizens." They also asserted that the passage of Heth's charter would open Virginia's coal trade to chartered companies, "the members of which may reside out of the state, and who by their vested rights and personal irresponsibility cannot be liable to the ordinary Legislation of the State."[30] A few weeks later, as Heth's charter worked its way though the Virginia legislature, the opponents of the act petitioned again. "No matter how guarded the Language by which such powers are conveyed, or the purity of intention with which they are asked or granted," they warned, "we have abundant evidence that they are frequently perverted, and made subservient to purposes not contemplated by those who were instrumental in their Creation." The miners also noted that only one corporation mined anthracite in Pennsylvania and that for several years the legislature "has been taken up in curtailing its powers, and counteracting its influence; and have therefore repeatedly refused to charter any other company."[31]

Despite such fears, mining corporations did not substantially alter the competitive structure of the Richmond basin. The provisions in its early mining charters ruled out expansionary strategies. For example, the charter of the Black Heath Company of Colliers made no provisions for expanding the firm and only incorporated the property already held by John Heth and his partners, Beverley Randolph and Beverley Heth. In fact, the charter prohibited the firm from purchasing any additional coal lands. In 1834 Abraham Wooldridge applied for a charter so that he could form a mining firm that could survive the death of one or more partners. The charter of the Cold Brook Company of Colliers (1835) contained a restriction on purchasing additional lands and divided the initial shares among the women and children of the Cunliffe family. In that case the corporate charter served as a way to consolidate a number of coal mining tracts tied to dower or inheritance rights. Elizabeth Branch formed two corporations, the Dutoy and Powhatan Coal companies, with her lessees on two separate tracts of coal land, thus uniting both parties of the lease under a single corporate concern. In the Richmond basin charters were rarely used to create completely new business ventures with the privilege to purchase or hold large amounts of land. As a result, the competitive structure of Virginia's established coal industry in the Richmond basin remained unaltered by corporate chartering.[32]

The increase in chartering in Virginia, however, did spur interest in some sort of reform of the process during the 1830s. In 1834 the General Assembly passed twenty-seven mining and manufacturing charters and a year later passed twenty-three more. These new industrial charters, coupled with the usual flurry of turnpike, canal, and bank charters, raised concerns among many Virginians that their legislature was creating too many corporations, with dire consequences. In 1837 the *Richmond Enquirer,* the leading Democratic newspaper in Virginia, sent out inquiries to New York, Massachusetts, and Maine regarding corporate chartering, in order to gain a comparative perspective on the process. The *Enquirer* reported that these states all placed significant restraints upon charters and suggested that "the true conservative doctrine be recognized on the face of the charter." In other words, the General Assembly needed to place restrictions upon the stream of charters passing through Richmond. A month later the Senate spent several days discussing a proposition to tax the stock of all incorporated companies, which drew the ire of pro-charter forces in the legislature. The *Richmond Whig* commented: "Somebody has very properly suggested that the title of such a bill should be—'An act to prevent the advance of Virginia in the arts and improvements of the age, and to send her [on] the downward road.'"[33]

Legal problems resulting from the increased presence of corporate charters in Virginia also signaled the need for reform. On a national level the status of corporations as "an artificial being, invisible, intangible, and existing only in contemplation of law," had been established by Chief Justice John Marshall in *Dartmouth College v. Woodward* (1819). This pivotal Supreme Court decision, however, served only as a general guide for dealing with corporate rights. The particulars of corporate rights and privileges were left to state courts. In Virginia, where private corporations were a relatively rare phenomenon before the 1830s, existing statutes left significant questions unanswered. In 1836 the Virginia General Court heard a case in which an individual tried to collect debts against a Lynchburg company whose charter had expired. Should the incorporators be liable for debts even though their firm no longer existed in the eyes of the commonwealth? In *Rider vs. the Nelson and*

Albemarle Union Factory Judge Henry Saint George Tucker argued that, according to the parameters of Virginia law, the managers of the company were no longer liable for the outstanding debt. He also postulated that the law states that, once a charter's life expires, all of its property reverts back to the commonwealth. Judge Tucker found both of these developments untenable and recommended a "general legislative provision" on the legal status of corporations as soon as possible. Thus, in order to clarify the status of corporations in Virginia's legal system, some sort of alteration in the chartering process was necessary. The rapid increase in chartering activity over the course of the 1830s made this reform all the more urgent.[34]

In 1837 pro-charter legislators introduced a bill that standardized the structure of business corporations and streamlined the chartering process. This legislation created a boilerplate charter that set stock subscriptions, the election of board members, remedies for delinquent stockholders, and other procedural rules. The bill also included a provision that a company's charter would be forfeited if less than five people ever owned four-fifths or more of its stock, if one person held over half of the stock, or if the company suspended operations for a period of two years. Manufacturing and mining charters were to last no longer than thirty years, and the legislature reserved the right to amend or repeal the charter after a period of fifteen years. The Democrats who introduced the bill in the General Assembly hoped that a boilerplate charter would reconcile the continued demand for corporations with the reservations many Virginians had about their passage. Despite efforts by Extra Billy Smith and other conservative senators to kill the measure, the act outlining general regulations for manufacturing and mining firms finally passed on 13 February 1837 and became law.[35]

Unlike Pennsylvania colliers, miners in the Richmond basin encountered few impediments to incorporation. Yet, even though they were relatively easy to secure in Virginia, the early use of charters by firms in the Richmond basin suggests that mining interests did not view corporations as much of a threat, as their brethren in the anthracite fields did. In Pennsylvania coal mining companies could secure charters but at a great political cost. In Virginia the legislature streamlined the process to create uniform charters that defused any potential battles between coal interests but also removed incentives to manipulate the system. As a result, more coal mining charters passed through Richmond than Harrisburg, but the creation of these corporations had a smaller impact upon the structure of Virginia's coal industry. As coal mining became more and more expensive in the Richmond basin, however, and as western miners sought to develop their extensive bituminous fields, the need for corporate organization became more urgent in the Old Dominion. By the 1850s both states revised their corporate chartering policies, with decidedly different outcomes for their coal trades.

A GENERAL OR A LOCAL SOLUTION

As the controversy over banks and corporations waned during the 1840s and 1850s, incentives grew for colliers of both Virginia and Pennsylvania to increase production. Throughout the 1840s national demand for coal increased steadily with the growth of household and industrial markets for mineral fuel in eastern cities. Furthermore, the new cities of the West continued to increase in population and manufacturing capacity, creating entirely new markets for Virginia and Pennsylvania coal in the Ohio and Mississippi valleys. Figure 5.1 illustrates the national upswing in bituminous coal production from the mid-1830s through the 1850s. Much of this coal was sent to western urban centers, for both domestic and industrial purposes. Bituminous coal shipments to Cincinnati reached 104,000 tons during the winter of 1846–47. Seven years later this figure had more than tripled, and by 1859–60 Cincinnati alone consumed 704,000 tons of bituminous coal. The appearance of "coal famines" in many cities during the 1840s and 1850s, moreover, signaled the need for expanded coal production and made speculation in the coal trade all the more lucrative for prospective capitalists. As the nation grew, so did its appetite for mineral fuel.[36]

At the same time, the cost of extracting coal increased in many mining regions. This was especially the case in Pennsylvania's anthracite fields, where many rich veins remained beyond the reach of small-scale mining enterprises. As coal lying close to the surface became scarcer, the entry cost for mining ventures became more and more expensive. During the 1840s the cost of mining below the water level was estimated to be five times that of mining above it, and the use of steam engines to hoist coal and pump slack water from mines increased considerably in the anthracite fields. By 1850 mines in Schuylkill

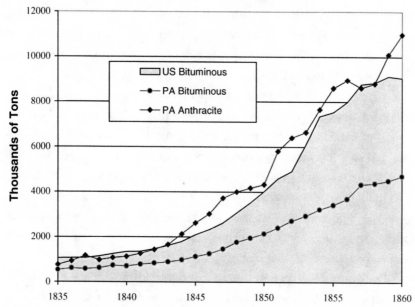

Fig. 5.1. National and Pennsylvania coal production in tons, 1835–1860.
(Eavenson, *American Coal Industry,* 430–31; *Historical Statistics of the United States: Colonial Times to 1970* [Washington, D.C.: Government Printing Office, 1975], 1:590.)

County alone employed 169 engines of nearly 5,000 horsepower, a number that nearly doubled by 1855. The bituminous mines of Pennsylvania's western counties rarely required extensive shafts or pumping engines in the 1850s, but the expense of connecting mines to major transportation lines served as a critical barrier to many individual entrepreneurs.[37]

The increasing demand for coal, coupled with the increasing cost of extraction as veins dipped beneath the waterline, triggered a widespread belief among coal interests that larger amounts of capital and land would be necessary to make a profit. Many states, most notably New York, offered "general incorporation," in which the executive branch issued charters to applicants that met administrative guidelines established by the legislature. This method of granting charters sought to remove incorporation from the legislature and formalize it into a more bureaucratic process. In 1848 Eli Bowen, a geologist and anthracite trade booster from Schuylkill County, bemoaned the lack of stability caused by Pennsylvania's chartering regime. Much of the instability, he concluded, could be attributed to the undemocratic ways in which firms received charters. "Their first step is to obtain *privileges* from the Legislature, which are not extended *generally* over the whole people," he argued. But this ran counter to Bowen's political assumptions, in which "many persons dispute the right of government to extend privileges to any class of men—believing that our Republican system should act by general, not by partial laws." General incorporation, therefore, could democratize the increasingly corrupt corporate chartering process. Even the militantly anticorporate *Miners' Journal* championed general charters and hoped that such an act would be passed, so that "it will place all of our citizens on an equal footing" and "form manufacturing companies, in which every one who is possessed of means,—and it will not require much,—will be permitted to join."[38]

Pennsylvania's first comprehensive general incorporation law, passed in 1849, applied to a number of industries. This act provided standard by-laws for the firm's operation, fixed a firm's minimum capital at twenty-five thousand dollars, allowed firms to own up to two thousand acres, and assessed a one-time tax of one-half of 1 percent of the capital stock.[39] The 1849 act targeted the manufacture of woolen, cotton, flax, silk, iron, paper, lumber, and salt but expanded over the next eleven years to include mineral waters, oil and paints, manures, marble, flour, "hydro-carbon fluids," leather, and peat. An 1853 supplement to the 1849 act extended its privileges to the coal industry as long as firms held less than two thousand acres of land, limited their capital to less than five hundred thousand dollars, and did not operate in the developing anthracite counties. The ban on incorporations in Luzerne, Northumberland, Lehigh, and Northampton counties was lifted the following year, but the restriction on Schuylkill County, still the center of anti-charter rhetoric in the anthracite region, remained in effect. This legislation eventually authorized the formation of 198 firms with more than thirteen million dollars in capital during its twenty-five year existence, of which seventy-eight were mining corporations. Coal mining firms, however, did not seem overly anxious to utilize the new general charter. Instead, they continued to turn to the legislature for special charters. In fact, soon after the legislature extended its privileges to coal mining firms, Governor William Bigler threatened to veto every special act passed for counties that were eligible for general incorporation. "Although this law is as favorable in its terms as the special acts solicited," he complained, "but a single application has been made under it for mining purposes."[40]

The Pennsylvania legislature passed another general incorporation act aimed specifically at opening up coal regions during the session of 1854. This permitted the incorporation of firms "for the purpose of developing and improving such mineral lands" but not for mining itself. A general charter from the 1854 act authorized the construction of railroads, machines, and structures for extracting the coal and even permitted firms to open shafts. It provided for the ownership of up to three thousand acres of mineral land and placed no limits on capitalization. The act also provided, however, that these firms "shall not engage, in any manner, in the business of mining, selling, or conveying to market the minerals on or in its lands." Two years later this limitation was repealed by a supplement that also placed a cap on capital of five hundred thousand dollars and prohibited the act from applying to Northumberland County. A year later the provisions of the act were extended to Schuylkill County, and in 1862 firms chartered under the 1854 act could increase their capital to one million dollars. Table 5.1 summarizes the provisions of Pennsylvania's antebellum general incorporation acts.[41]

Table 5.1. Antebellum General Incorporation Acts of Pennsylvania

	PA 1849 (1853 Supp.)	PA 1854
Number of coal companies created through 1860	73	30
Limit on capital	$500,000	No limits to 1856 $500,000 to 1862 $1 million after 1862
Limit on size	3,000 acres	3,000 acres
Railroad rights?	Yes—on own lands	None
Excluded counties?	Luzerne (to 1854) Northumberland (to 1854) Lehigh (to 1854) Northampton (to 1854) Schuylkill	Northumberland (after 1856) Schuylkill (to 1857) Northampton Lehigh, York
Liability	Total liability (to 1854) Labor, machinery, and materials Debts (after 1854)	Labor and materials Total liability (after 1856) Limited liability (1860)

Although it would seem that Pennsylvania's general laws might supplant special charters, for many colliers the 1854 law was not satisfactory. Governor James Pollack himself, in his message to the Assembly of 1857, complained that the laws regarding manufacturing and improvement companies were too strict and did not attract sufficient investment. He argued that, "instead of encouraging individual and associated enterprise and energy in the development of our immense natural resources, they bind and crush both by severe restrictions, unwise limitations, and personal liabilities." Even after the passage of these general acts, moreover, the legislature remained the locus of chartering activity in Pennsylvania. In 1855 the *Miners' Journal* noted that "almost every other bill reported in the Legislature is a supplement to some Coal or Iron company, in this State, to obtain privileges for their neighbors." But, as the years progressed, wouldn't the presence of general laws replace the special charter? After all, companies that had incorporated under a special charter still had to negotiate the rapids of the legislature in order to get amendments to their charter to raise capital, increase the amount of land they could own, and even change the name of the firm. The general laws definitely made an impact upon corporate chartering in the coal industry, but they did not replace legislative charters.[42]

So, why did firms continue to choose special incorporation, with its treacherous prospects, when general charters could be easily had? The answer lies in the fact that a successful special charter often provided a bonanza of privileges, such as the authorization of massive capitalization, generous land grants, or, best of all, a vaguely worded section that contained no specific guidelines so that the firm itself could decide what was proper. Many firms published their charter along with their prospectus in order to convince potential investors that the legislature, in addition to the market, had provided a "can't miss" opportunity to make money in the coal industry. Thus, while special chartering was by no means a guarantee, the payoff often justified the effort. Contemporary observers struggled to find rhyme or reason in Pennsylvania's system of corporate chartering, but the complicated mix of general and special charters actually resembled a de facto industrial policy of coal in Pennsylvania. This policy effectively targeted specific regions for development and protected existing interests from competing charters.[43]

The 1849 act's 1853 amendment that allowed coal mining firms to receive general charters, for example, proved particularly effective in attracting capital investment in the northern anthracite field. Of the seventy-eight firms chartered under the act, sixty-two (80%) were authorized to own lands in Luzerne County, and the vast majority of the firms that incorporated did so in the five-year period from 1854 to 1859. This concentration of firms in the northern anthracite field most likely developed in response to the needs of out-of-state investors. While the southern and middle fields were linked to Philadelphia, the northern field shipped mostly to markets out of state via interstate carriers such as the Delaware and Hudson Canal. The most obvious destination for anthracite coal mined in the northern field was New York City, not Philadelphia.[44]

New Yorkers, however, remained justifiably skittish about investing in Pennsylvania coal lands without a Pennsylvania charter—in 1833 the state had forced two New York corporations, the Delaware Coal Company and the North American Coal Company, to secure state charters or lose their lands. Rather than deal with the political mischief often required to get special charters, many New York investors chose instead to use the 1849 general incorporation act, once it had been amended to issue mining charters. Despite a clause that mandated that the majority of the individual stockholders be residents of Pennsylvania, many of these firms were obviously based in New York and merely retained a majority of Pennsylvania stockholders as figureheads. For example, the Luzerne Anthracite Company's minority directors, Samuel Smith and George Haywell of New York, owned 98.5 percent of the stock. Smith also owned 98 percent of the stock of the National Anthracite Coal Company, 99.5 percent of the East Scranton Coal Company's stock, and 88 percent of the North Carbondale Coal Company's shares. One trade journal in 1856 noted the heavy development of the northern anthracite field, mentioning that the purchasers of coal land "to facilitate their enterprises, have very generally concentrated their capital in companies formed and registered under the General Mining Law." In the end, however, corporate development of the northern field probably had positive results for the overall anthracite trade, as over the five-year period from 1855 to 1860 coal production there rose 64 percent, as opposed to 40 percent for the middle field and 4 percent for the southern field.[45]

Whereas the 1849 act served to concentrate investment in the northern field, the 168 mining corporations created under the 1854 act were spread all over the state. Schuylkill County, the bastion of individual enterprise, led the way with 37 percent of the charters, Luzerne County had 21 percent, Northumberland County in the Middle Anthracite Field had 7 percent, and bituminous Westmoreland County had 5 percent, with the rest spread out among some twenty counties. All in all, the 1854 act authorized firms to raise nearly eighty million dollars in capital and opened over 130,000 acres to development.[46]

Although the 1854 act did not seem to favor any particular region or county, it did address the problem of the improvement companies, whose broad corporate privileges resembled a real threat to existing coal firms throughout the 1850s. Improvement companies were firms authorized by the legislature to do everything *except* actually mine coal. They could own land, build structures and railroads, find coal veins, and sink shafts, but they had to leave the extraction of minerals to other firms or individuals. Yet, despite these limitations, many politicians and small colliers viewed improvement companies as "horrible monsters" whose limited liability, large capitalization, and ability to hold and lease large tracts of land threatened individual enterprise.[47]

The 1854 act standardized improvement company charters—thus reducing popular fears that these firms would overwhelm coal regions—but at the same time encouraged the corporate development of mineral lands. As special acts of the legislature, improvement company charters were notoriously open-ended. Such pliant charters carried virtually no restrictions on acreage, capitalization, or location. General incorporation, however, standardized all of these provisions by creating a boilerplate charter. In a sense the 1854 general act leveled the playing field and effectively undermined much of the appeal of improvement companies. Thus, rather than eliminate the improvement company altogether, general incorporation in this case represented a way of legitimizing improvement company charters while protecting existing coal interests.[48]

General charters thus targeted specific areas for specific types of development by placing certain counties off limits to general charters while keeping others open to special charters. A closer look at the special charters passed after the rise of general incorporation in coal mining demonstrates that Pennsylvania's legislative charters followed this geographical pattern. More specifically, the legislature tended to create western bituminous corporations with railroad privileges during the 1850s. Of the twenty-seven coal and improvement companies chartered by the legislature from 1855 to 1860, twenty (74%) were based in the western bituminous field, and eleven (41%) were western bituminous firms specifically granted the power to build railroads to link coal lands with existing lines. Comparatively, the general acts tended to be eastern firms; out of the sixty-five charters granted by the 1849 act over the same period, 88 percent mined anthracite coal in eastern Pennsylvania. Sixty-eight percent of the twenty-five charters issued under the 1854 act were authorized to mine eastern anthracite coal. Less-developed regions could count on fewer enemies in the legislature, and, as long as prospective firms did not encroach upon an existing firm's interests, their charters had a better chance of passing. The simultaneous existence of general and special charters therefore opened both the existing anthracite fields as well as the growing bituminous fields of Pennsylvania to corporate development at the same time that they reduced the political conflict over corporate chartering.[49]

General incorporation did not "depoliticize" the corporation in Pennsylvania. Instead, it represented one option for corporate growth in an incredibly complex, but ultimately successful, institutional framework that prioritized rapid economic development. The use of chartering policy to increase production directly was not a new idea in Pennsylvania. In 1827 the *Miners' Journal* suggested that coal company charters should contain a provision that compelled them, on the pain of losing their charter, to mine five thousand tons of coal in their first year of operation, eight thousand in the second, twelve thousand in the third, and fifteen thousand every year after that. The 1836 general act issuing charters to firms that made iron with anthracite coal is another example of chartering policy targeting specific goals. In this light chartering reform was not a simple process of expanding the benefits of incorporation to a wider population of entrepreneurs. State officials used corporate charters, instead, to target particular areas for development, to limit ruinous competition in others, and to remove some of the nastier elements of special incorporation from the legislature. Rather than remove politics from chartering, the appearance of general incorporation in Pennsylvania infused even more political considerations into corporate reorganization of the coal trade.[50]

During the 1840s and 1850s Virginia's coal trade also witnessed a growing demand for the larger capital and limited risk of the corporate charter. In the Richmond basin much of the easy coal had been won by the advent of the 1840s, which increased the expense of raising the coal for colliers. Meanwhile, the population increase in the western counties opened the bituminous fields of the Kanawha and Ohio to potential development. As figure 5.2 demonstrates, coal production in the Richmond basin remained stable or declined at the same time that western Virginia colliers increased production.

Thus, as in Pennsylvania, the technological reorganization of existing fields through the use of deeper shafts, more efficient water pumps, ventilation systems, and other innovations, plus the added cost of purchasing or hiring slave labor, made the investment of more capital in the Richmond basin mines much more necessary than in the past. Virginia's western fields, with

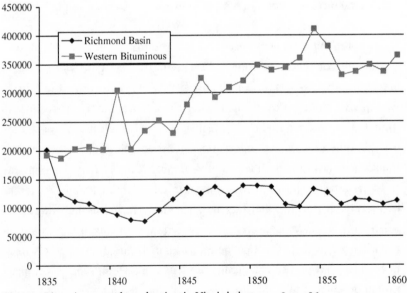

Fig. 5.2. Bituminous coal production in Virginia in tons, 1835–1860.
(Eavenson, *American Coal Industry,* 508.)

their need for quality surveys, transportation links, and other expensive start-up costs, also required corporate development. As with the earlier challenge to corporate chartering, the solons at Richmond again altered the state's chartering regime. Much like internal improvements, geological surveys, and just about everything else pertaining to Virginia's coal industry, corporate chartering had become embroiled in the Old Dominion's sectional debate over slavery and representation during this period. Virginians responded by developing a local solution to the problem of corporate chartering during the 1850s, which most likely hindered the growth of the coal trade of the western counties.[51]

Western counties only occasionally participated in the Jacksonian debate over charters. The most striking example occurred in 1835, when a group of salt manufacturers of the Kanawha Valley argued that "the property & capital of all the individuals engaged in the manufacture of salt is gradually but inevitably wasting away," and they petitioned the legislature for a charter of incorporation. Lukewarm support by Kanawha County politicians in Richmond combined with local opposition to kill the idea. "The whole Kanawha Valley will be tributary to this Company," a counter-petition to the salt manufacturers claimed, "and it will exercise its power with less clemency because that power will be as perpetual as the perpetuity of its charter." Anti-charter forces in Kanawha County gradually broke down as they realized that larger amounts of capital were needed to compete with the salt-producing regions of New York and Ohio. In 1847 salt manufacturers formed the Kanawha Salt Association to oversee production and regulate prices. In a way anti-charter suspicions were correct. By 1851 a single joint-stock company controlled all but one saline manufactory in Kanawha County.[52]

The Virginia legislature did not issue any coal mining charters in the trans-Allegheny region until the 1840s. The Preston Railroad, Lumber and Mining Company of Preston County was created in 1840 with the authority to own ten thousand acres. In 1848 the West Virginia Coal Mining Company received a charter from the General Assembly with the authority to hold ten thousand acres in six counties and was capitalized at one million dollars. Although many of these companies never actually brought coal to market, chartering activity in western Virginia suggests a growing interest among investors in the area's mineral potential. The discovery of cannel coal, which could be easily distilled into a coal oil or gas for illumination, east of Charleston in the Kanawha Valley provoked even more interest in western Virginia among followers of the coal trade. In 1851 *Scientific American* reported that "we have not hesitation in saying we believe it to be equal to any coal of the same kind in the world" and that "we trust that what we have said may be the means of bringing it more prominently into public notice." By the early 1850s the mineral wealth of Virginia's western counties had begun to attract serious attention from investors across the United States and even in Great Britain.[53]

Western Virginia's coal trade drew national interest by the late 1840s and early 1850s, but political developments in Richmond threatened to squelch the area's prospects. As noted earlier, the tension between western and eastern Virginians triggered a constitutional convention in the fall of 1850 to redress western grievances regarding legislative apportionment and taxation. Corporate chartering was not a major topic in the convention of 1850–51, but the existing chartering policy in Virginia became a casualty of one of the new constitution's reforms. In the interests of limiting the number of laws required to enact divorces, pensions, and other specialized issues, the delegates nearly unanimously agreed to limit the legislature to biennial sessions of ninety days.[54]

The 1850 convention alleviated many of the west's concerns over suffrage and representation by granting universal white male suffrage and promising a future reapportionment of the legislature, but it did not completely dispel Virginia's sectional strife. In fact, legislation continued to take the zero-sum approach to an even greater extent during the 1850s. Sectional tension, especially in the areas of internal improvement, continued to dominate legislative policy in Virginia. "The bane of our Legislation is sectionalism, of the most bigoted and narrow kind," Richmond's *Daily Dispatch* argued. "Each man, wrapped up in his own local schemes, has no eye for a comprehensive view of the general interests of the State." After the Virginia legislature began to meet on a biennial, rather than annual, basis, this statement seemed only a slight exaggeration. But biennial legislative policy had little real consequence for the agricultural interests of the state. Most major projects in the tidewater were either finished or close to completion, so the slow pace of legislative activity was of no real concern. Even the James River and Kanawha Canal, long the whipping boy of dissatisfied farmers, stretched over 196 miles to Buchanan in the Blue Ridge Mountains by 1851. This might account for the strong support of eastern delegates for the biennial session reform enacted by Virginia's new constitution.[55]

For other Virginians the lack of an annual session created major problems. Most notably, demands on the legislature for corporate charters occupied a large portion of the session. "The State is now controlled by corporations," argued tidewater senator John W. C. Catlett in 1854. "The heads, the Agents, and sub-agents of corporations, have crowded this chamber and the other all winter." During the relatively leisured pace of Virginia's annual session, charters often required intensive lobbying efforts to ensure passage. A biennial session shortened the time horizon for legislative lobbying. With only ninety days to secure a charter and with the prospect of waiting two years if they were unsuccessful, it seems logical that corporate lobbies would turn up the intensity in a biennial session. The frenzied pace, however, did not translate into more opportunities for western colliers. "As to expecting the Legislature to do anything for us this winter it is leaning on a broken reed," exclaimed one Kanawha Valley industrialist in 1855. "The set of miserable wretches! They will spend their own time & the people's money discussing federal politics, the Crimean War, they may also frighten the old Banks out of their wits, (I wish they would break them up)." For Virginians interested in developing the mineral resources of their state, the biennial legislature was not a helpful change.[56]

During the early 1850s, therefore, development-minded members of the General Assembly sought to reform chartering policy to mesh with the newly overhauled state political structure. The natural choice, given the experience of states such as New York, would be to pass some sort of general incorporation law that transferred chartering authority to the executive branch. Throughout the antebellum period, however, Virginia's governor held little authority. The executive branch had never exercised any veto power over the General Assembly, and the constitution of 1850–51 did little to enhance the governor's power within Virginia's state government. Aside from pardons and militia commissions, the executive branch had very limited capacity and could not assume the responsibility of general incorporation in the 1850s.[57]

Instead, Virginia chose a local solution to the problem. As early as 1852, the legislature considered plans for using circuit courts, the basic county- and city-level judicial unit in Virginia, to incorporate savings banks. During the same session Albert Reger, a senator from western Virginia, suggested that circuit courts should also be empowered to incorporate all agricultural and manufacturing companies. At the next session Spicer Patrick, a delegate from Kanawha County, introduced a bill to use circuit courts to incorporate mining and manufacturing companies. Under Patrick's guidance the bill was passed into law on 3 March 1854. In addition to allowing circuit court judges to issue charters, this act limited the liability of stockholders up to the total of their investment and provided that each corporation created in this fashion must open its books on the legislature's demand. The act limited the amount of land that coal mining companies could hold to 3,000 acres and mandated capitalization at no less than $20,000 and no more than $100,000. Coal companies chartered by the legislature in 1854 averaged $818,181 in maximum capitalization and 3,909 acres, so circuit court charters would be considerably smaller in size than their legislative counterparts.[58]

This act, like nearly every issue in antebellum Virginia politics, had a sectional dimension. Western interests spearheaded the charge for circuit court chartering for mining and manufacturing companies. Entrepreneurs from western Virginia still considered themselves second-class citizens in Richmond, and securing a charter during an abbreviated session complicated matters for potential coal mining corporations. The roll call of the vote in the Senate suggests a strong sectional approach to the corporate chartering bill. Of the twenty-three yea votes, a quarter came from eastern senators, while three-quarters came from the western counties. The nays were decided along the same lines.[59]

Why circuit courts? As mentioned earlier, Virginia lacked administrative capacity at the executive level. State authorities could have easily established new bureaucratic agencies to supervise chartering, but, instead, Virginians decided to rely upon established local institutions. The long-standing tension between state and local authority most likely played a large role in determining the localist flavor of Virginia's chartering reform in the 1850s. Since the eighteenth century Virginia society had a well-established "Country" tradition that remained suspicious of centralized, or "Court," directives. In his study of Virginia legal culture in the eighteenth and early nineteenth century, A. G. Roeber claimed that Virginians identified the ideas of manufacturers, bankers, and many reformers as a Court mentality that should always be held in suspicion. During the 1850s this traditional distaste for Court solutions to legal problems blended easily with the growing sectional tension over slavery to rule out following the leadership of New York, Massachusetts, and Pennsylvania in corporate chartering. "The law, whether of a constitutional or civil variety," he argued, "could be regarded as legitimate for many southerners only if they felt it to be close to their way of life and to their values." Indeed, Virginia's entire legal system reinforced an agricultural and decentralized vision of society. This system, like most political institutions in Virginia, shielded slavery and ensured that oligarchic structures persisted in the Old Dominion throughout the antebellum era.[60]

Western Virginians interested in economic development chafed under the rule of county courts, but, as the minority, they were forced to work within the decentralized framework of Virginia politics. Because Virginia's legal system had preserved the rights and interests of large landholders through propertied suffrage, farmers still dominated many county-level institutions in the west—much to the chagrin of development-minded residents. Waitman Willey, a delegate from western Monongalia County, argued on the floor of the 1850–51 convention that "there are more assaults and batteries perpetuated sometimes, than there are new friends formed or old ones strengthened at these courts." Not only do county courts "have a demoralizing influence, rather than an educational bearing," Willey maintained, but they continued "disturbing the industrial pursuits of the people, and often promoting discord in their social relations."[61] Despite resentment from a vocal minority, the majority of legislators regarded county-level courts as the appropriate locus of chartering authority in Virginia. Considering that farmers constituted 55 percent and lawyers only 17 percent of Virginia's House of Delegates in 1854, this Country reform made sense for many legislators. The new constitution also provided for the popular election of circuit court judges, which removed the position from Richmond's influence. Institutionally and culturally, circuit courts were a logical location for chartering authority.[62]

Table 5.2. Sectional Breakdown of Virginia Circuit Court Charters, 1855–1860

Total Charters	East/West Total	East/West Industrial	Total Coal	Western Coal
69	37/32	22/22	25	21

Sources: "Communication from the Secretary of the Commonwealth, Enclosing List of Compa-
nies Incorporated by the Courts of this State," Senate doc. 2: *Virginia Senate Documents, 1857–58*
(Richmond: John Warrock, 1858); "A List of Mining and Other Companies, Incorporated by the
Circuit and County Courts," doc. 25: *Virginia Legislative Documents, 1859–60*, vol. 6 (Richmond:
John Warrock, 1860); "A List of Mining and Other Companies, Incorporated by the Circuit and
County Courts," doc. 26: *Virginia Legislative Documents, 1861–62* (Richmond: n.p., 1862).

Although circuit court charters offered a political solution, they did not
necessarily aid in the development in Virginia's western coal trade. As table 5.2
demonstrates, most of the coal charters were issued by western courts, but
eastern courts issued the most charters overall. This trend can be explained by
the tendency for eastern courts to charter nonindustrial firms such as tele-
graph companies (6), hotel companies (5), libraries (3), colleges (3), cemeter-
ies (2) and miscellaneous firms such as real estate companies and resorts. The
number of eastern and western industrial charters is actually even. The impact for coal companies is even more striking. Of the twenty-five companies
chartered by circuit court judges from 1855 to 1860 with the privilege of mining coal, twenty-one (84%) of them received their charters in the West.

Nevertheless, relocating chartering authority to the county level hampered, rather than helped, the quest for corporate capital in western Virginia.
This move provided access to general charters but with different results than
Pennsylvania's de facto industrial policy of the 1850s. Pennsylvania's experience, especially in the northern anthracite field, proved that capital needed to
flow smoothly across state boundaries by the eve of the Civil War. Wary investors required assurances that their money was well spent, and the provisions of mining charters came under close scrutiny during the 1850s. In 1854
New York's *Mining Magazine* described the potential pitfalls of investing capital in mining ventures. "Another embarrassment to legitimate mining has
arisen from a defect in the legislation of the States," the editors argued, "or
from a neglect to enforce the provisions of the laws enacted." Constant vigi-
lance, they advocated, was the only way to avoid fraudulent mining corporations.[63]

Simply put, getting a corporate charter was not enough to attract capital investment by the 1850s. Pennsylvania's myriad of general and special charters offered a number of options to coal mining firms. In Virginia, however, these options were not as attractive. Prospective corporations in the Old Dominion had to risk the biennial legislature or get a circuit court charter that limited capital to $100,000. By 1855 many New York investors considered $1 million capitalization the minimum for mining ventures, so a circuit court charter appeared woefully inadequate to attract capital. In 1854 the Virginia legislature amended circuit court chartering to increase maximum capital to $400,000. Kanawha County's Saint George Mining and Manufacturing Company nonetheless received a charter in 1855 which set capital at $1.5 million. The idea that judges had such authority to create (and potentially deny) such charters was not the kind of stability that the *Mining Magazine* desired. What would happen if a rival company challenged the legality of the St. George Company's charter? Were firms operating with circuit court charters in Kanawha County released from limits upon capitalization? What if the firm wanted to expand beyond Kanawha County? Would the directors need to secure a charter from the neighboring county? Virginia's combination of biennial and local chartering might have resulted in easier incorporation, but it did not reinforce the validity of those firms to outside capital.[64]

Virginia's chartering policy also impeded investment in the western bituminous fields by making it especially difficult for outside investors to get information regarding their existence. "It is not the merits of the speculation itself, but the want of capital, through the unfounded aversion of the public, that is the chief cause of failure" to attract investment, argued the *Mining and Statistic Magazine* in 1858. Capital was available for coal companies, the editors asserted, "but in the case of Mine shares there is generally a difficulty in finding a purchaser, except among those intimately acquainted with the merits of the concern, arising from the prevailing antipathy to such undertakings, and the general want of information respecting them."[65] In Pennsylvania Harrisburg acted as the clearinghouse for both special and general coal company charters. New York, Boston, and Philadelphia investors could draw upon the vast network of professional lobbyists, lawyers, and former legislators to guarantee that they received the right kind of charter. Economists might characterize this relationship as an "external economy of scale"—in which factors outside the influence of a single firm decrease the average transaction costs of receiving a charter. Institutions such as the Kanawha County circuit court, although an important legal and social center for the local area, could not develop a similar network. Local entrepreneurs tried to promote their endeavors through prospectuses, but, without the high profile of Pennsylvania coal, they operated at a disadvantage. Thus, placing chartering jurisdiction in the hands of local authorities chilled investment by keeping information costs high for potential investors at the same time that western coal needed the most promotion.[66]

To make matters worse, chartering authority overlapped in Virginia. Unlike in Pennsylvania, where certain counties remained off-limits to general charters and informal constraints checked the promulgation of special charters in other areas, Virginia placed no such geographical restrictions on either legislative or general charters. This arrangement complicated corporate chartering in the west. Judges in western Virginia could not issue charters with wide-ranging railroad privileges or with the power to own land outside of their jurisdiction. Even if a firm did secure a circuit court charter, there was no guarantee that a rival could not secure a charter with the privilege of raising more capital and owning more land from the legislature. Of the thirteen circuit court charters from western Virginia which list capitalization amounts, the median maximum capitalization was $300,000, and only one conveyed the right to build railroads. From 1854 to 1860 thirty-five special charters passed the legislature for coal mining firms in western Virginia, with a median maximum capital of $500,000, landholding privileges of 5,000 acres, and nineteen (54%) of which could build railroads. Rather than complement corporate investment as they did in Pennsylvania, general and special charters competed with one another in western Virginia.[67]

Throughout the 1850s the western coal regions of Virginia continued to have trouble attracting capital investment from outside the area. In 1859, for example, the editors of the *United States Mining and Railroad Register* responded to a series of articles boosting western Virginia in the *London Mining Journal.* "It could thereby deduced or made to appear that the *Canawha* district in America comprised a promising field for British investors in mineral lands or colliery operations," the editors of the *Register* admitted. But this interest "was only repeating, in a more compact and better plan what had been attempted before, without success." British geologist D. T. Ansted remarked that the vast potential of the western Virginia bituminous fields could not overcome a dearth of capital in his 1854 travel narrative entitled *Scenery, Science, and Art; Being Extracts from the Note-Book of a Geologist and Mining Engineer.* "It is indeed utterly impossible that any such amount of capital as is at present engaged in the Western coal trade, or is likely to be engaged in it for some time to come," Ansted wrote, "can raise and convey the coals fast enough to the market to cause competition." Western coal mining remained capital starved on the eve of the Civil War. "A few more similar transactions," wrote an agent of the Paint Creek Coal Company in 1859, in response to a recent purchase of stock by Cincinnati investors, "would dispel the cloud that appears to dim the prospects of the valley as a field for profitable investment."[68]

The experience of western Virginia in the 1850s suggests that liberal chartering in and of itself only works in the right context. Pennsylvania's blend of special and general acts provided the right mix of vast privileges in undeveloped areas with rather strict restrictions in others. Thus, the unique blend of anticorporate ideology with a responsive legislature resulted in an effective industrial policy of development. Virginia's chartering regime potentially blended large speculative corporations in the legislature along with small accessible charters in circuit courts. Limiting special charters to biennial sessions and creating circuit court chartering in the remote areas of western Virginia, however, hardly emerged into a successful developmental policy. At a time when Virginia needed to advertise the opportunities in its western sections aggressively, the legislature shifted a great deal of policy-making responsibility to local authorities. By the 1850s New York, Philadelphia, and Boston capitalists knew the anthracite and bituminous fields of Pennsylvania. They did not, however, regard western Virginia as a particularly lucrative investment opportunity, in part because of the lack of adequate transportation and geographical knowledge but also because of chartering problems. The local solution seemed consistent with the Old Dominion's past but woefully inadequate for Virginia's future.

The corporation represented a critical intersection of public policy and private enterprise during the antebellum era. Both the Virginia and Pennsylvania legislatures struggled to reconcile the demands for corporate development of their coalfields with long-standing reservations about the impact of corporations upon society. In the end Virginia turned to its tradition of decentralized governance in chartering policy, whereas Pennsylvania expanded chartering options. As with internal improvements and geological surveys, the creation of chartering regimes in Virginia reflected the adherence of most policy makers to a conservative and primarily agrarian political economy. Pennsylvania's chartering policy, like its experience with the State Works and the geological survey, revealed both the strengths and weaknesses of Pennsylvania's responsive political system. The interplay of political interests and economic aspirations translated into a system of granting charters which allowed for rapid development of new mineral regions as well as the continued prosperity of existing fields.

The story of corporations and coal in Pennsylvania and Virginia stands as an important narrative of economic aspirations, political intrigue, and both intended and unexpected results. It also represented the last component in the antebellum political economies of coal for both states, as the advent of the Civil War induced tremendous changes in their institutional structures. In Pennsylvania the wartime and postwar legislature struggled to balance the interests of corporate actors, the rising and increasingly militant miners, and a public that relied more and more upon coal as a domestic and industrial fuel. In doing so, policy makers in Harrisburg moved toward a system that was even more dependent upon mining corporations. The Civil War solved Virginia's sectional squabble once and for all with the creation of West Virginia in 1863. The course of development pursued by the new state, however, turned out to be less than successful. As the next chapter demonstrates, the 1860s wrought major changes in the policy-making regimes of both Pennsylvania and the Virginias.

Three Separate Paths

The Impact of the Civil War

I n 1902 Granville Davisson Hall published a lengthy treatise on the creation of West Virginia entitled *The Rending of Virginia*. As an eyewit-
ness to the state's creation from 1861 to 1863, Hall underscored the long history of sectional divisiveness in the Old Dominion when he wrote that
"Eastern Virginia always possessed a full endowment of the selfishness and
blindness to its own true interests inherent in aristocratic communities." "The
aristocrat takes in only the little circle of which he deems himself the center,"
he argued, "regarding all outside of it alien and hostile." When it came to the
actual break between Virginia and West Virginia, Hall knew exactly what to
target as the divisive issue. "Like the watch which in spite of everybody persisted in keeping wrong time till the magnet secreted near the mainspring had
been discovered, nothing could go right in Virginia till the concealed but evil
influence of slavery had been removed," Hall asserted. "Other questions might

come and go, other causes be lost or won; but while this wrong remained as a basis of injustice and irritation, it would rankle and breed fresh dissension from year to year, until East and West should be rent asunder in fact as they had long been in feeling."[1]

Although they tend to use less inflammatory rhetoric, historians often characterize the Civil War as one of the great "modernizing" forces in U.S. political history. In this view the war swept aside the stumbling block of slavery, settled many of the nagging problems of the relationship between federal and state authority in the United States, and paved the way for a more active federal government. Like Hall, many historians have commented upon the long-standing grievances of non-slaveholders living in southern states, although West Virginia became the only region actually to sever the relationship between former slave and free labor regions. Once the issue of slavery became irrelevant in the years following emancipation, they argued, Americans no longer regarded their nation as a "house divided" between free and slave states. As the harbinger of the fast and loose political economy of the Gilded Age, the Civil War also marked the complete breakdown of state-level governance of the economy. A few quaint reminders of the antebellum "producerist" mentality lingered into the 1870s, but ultimately the postbellum decades saw the withering of regulatory authority among states and the emergence of a truly national market. The Civil War, therefore, became an economic and political watershed in the eyes of many contemporary and historical observers.[2]

Yet how clean a break did the Civil War create in the politics of industrialization? How could four years of war completely recalibrate the role of the state in the nation's political economy? The changes in the coal trade during these years suggest that the Civil War altered state-level institutional contexts, with critical implications for the developmental paths of Pennsylvania, Virginia, and West Virginia. But, rather than make state policies irrelevant or ineffectual, political developments during the war years reconfigured institutional frameworks to produce an altogether different environment for corporations, individual proprietors, and workers. An anticipated increase in the demand for both anthracite and bituminous coal spurred interest in Pennsylvania's eastern and western coal regions during wartime, for example, but the legislature struggled to meet the challenges of raising an army, mobilizing the economic resources of the state for the war, and maintaining domestic order amid the chaos of wartime. In order to manage this task, state officials provided for a massive corporate reorganization of Pennsylvania's coalfields and allowed large railroad corporations to coordinate the production and transportation of coal during wartime. Corporate tax revenues and production swelled to new heights in both the anthracite and bituminous regions as result of this new policy. But for small colliers and wage-earning miners the reliance upon large corporate interests to increase production and tax revenues was a Faustian bargain. In postwar Pennsylvania the same railroad corporations dominated the state's coal trade in a fashion that antebellum policy makers would have deemed intolerable.

The unique conditions of the Civil War blocked opportunities for individual colliers and workers in Pennsylvania, but it opened new doors for western Virginians. Amid the wreckage of military conflict, the creation of West Virginia in 1863 initially spurred optimism among colliers of the Kanawha and Ohio valleys. By 1865 Pottsville's *Miners' Journal* celebrated the liberation of the coal trade of the Kanawha Valley, which had been "cursed and controlled by the slave masters who 'like the dog in the manger,' have for fifty years denied it to enterprise, and knew not how to profit by its immense mineral wealth themselves."[3] Separate statehood did eliminate the influence of Richmond's conservative legislature over the region, but the new state suffered from a number of problems during its infancy. West Virginia's founding generation wanted the new state to resemble its northern neighbors, Ohio and Pennsylvania. "Radical" West Virginians therefore attempted to form new policy regimes to facilitate the simultaneous economic and social development of their state.

In the end all three states departed from the well-established patterns of the antebellum years and faced new challenges during the Civil War. The war years even offered a last hurrah for Richmond basin colliers—albeit one brought about by military necessities and not any long-standing changes in the technological or organizational structure of the region. The great conflict altered the structure of the coal trade in each state, as could be expected. It did not, however, end the role of the individual state in the U.S. coal trade. Nor did it help integrate each state's coalfields into a unified national economy.

Granville Davisson Hall described the Old Dominion as "rent asunder" as a result of the Civil War; he might have used the same words to describe the political economy of coal in Pennsylvania, Virginia, and West Virginia.

FUEL FOR THE UNION: THE CIVIL WAR
AND PENNSYLVANIA COAL

Pennsylvania politics, like that of most states, underwent a major partisan realignment as the sectional crisis in the federal government reached its crescendo during the presidential election of 1860. Over a twenty-year period beginning in the late 1830s, a relatively stable two-party system of Democrats and Whigs controlled Pennsylvania's state government. In 1854 the state's Whigs fell victim to internal squabbling over the slavery issue and suffered a rash of defections to the growing nativist movement. The Pennsylvania Democratic Party, only nominally led by a native son, President James Buchanan, likewise showed signs of decay brought about by years of coalition building and compromise with other political factions. During the late 1850s the newly formed "People's Party" united former Whigs, Anti-Masons, and Know-Nothings, as well as disgruntled Democrats, to reshape partisan politics in Pennsylvania. The People's Party adopted the free labor stance of the Republican Party, complemented it with a strong protectionist stance on tariffs, and sufficiently distanced itself from the abolitionist wing of the national Republicans to capture Pennsylvania's lower house in 1858. Andrew Curtain, the People's Party candidate for governor in 1860, won by a 32,000-vote margin and brought a new ruling party to Harrisburg.[4]

But a stormy relationship between the state and Pennsylvania's largest railroad corporation, the Pennsylvania Railroad (PRR), quickly eroded any mandate enjoyed by Pennsylvania's version of the Republican Party and undermined public confidence in the legislature. The Pennsylvania Railroad had paid a tonnage tax since its creation in 1846 to compensate for its competition with the State Works system. In 1857 the PRR agreed to purchase the main line of the State Works for $7.5 million in PRR bonds. The original agreement also provided for the abolishment of the PRR's tonnage tax and released the firm from any taxation in the future, but Pennsylvania's Supreme Court ruled the latter provision unconstitutional. At the time J. Edgar Thomson, the PRR's president, refused to grease the palms of unscrupulous legislators in order to rid his firm of the tonnage tax. Thomson stood firm in his conviction that "this company cannot procure the necessary legislation without resorting to means that all proper minded persons must condemn."[5]

But by 1860 Thomas Scott, the PRR's dynamic new vice president, had no such qualms. Scott placated disgruntled legislators from western Pennsylvania with the promise of new railroads in their region and uniform rates. He also called upon the services of influential Republicans such as Pennsylvania's Senator Simon Cameron to swing reluctant legislators into the PRR camp. When that failed to win enough votes, he used outright bribery to persuade political opponents to change their position on the tonnage tax. Scott and his Republican allies introduced the bill amid the secession crisis of 1861 and, after a narrow victory in the State Senate, a reluctant Governor Curtain signed the repeal of the tonnage tax into law in April 1861. Public opinion quickly turned ugly. Despite the distraction of secession and war, voters took their frustration out on incumbents in the subsequent state elections of 1861. Only one House member who voted for the repeal returned to office, and the only senators to survive the fallout were the two who had led the fight against the PRR.[6]

This controversy allowed the Democrats to recapture the House and increase their strength in the Senate by the 1862 session. Almost immediately, legislative committees formed to investigate the corrupt practices used to secure passage of the PRR's bill. New legislation to reinstate the tonnage tax appeared on the floors of the House and Senate. The PRR ordeal shook the electorate's confidence in the Pennsylvania legislature and made legislators justifiably reluctant to take public stands on issues of any consequence, lest they suffer the wrath of angry constituents. Party discipline during the war years eroded as the People's Republican, Republican-Union, Democratic, Constitutional-Union, and other organizations vied for control of the legislature. Local legislation, with its propensity for vote trading, last-minute introduction of bills, and underhanded passage in the closing minutes of the session therefore remained the main occupation of the Pennsylvania legislature through much of the 1860s.[7]

Amid partisan and electoral unrest, Pennsylvania's state government shouldered the burden of raising and equipping a modern army within a muddled and constantly changing federal-state military system. Following the surrender of Fort Sumter on 14 April 1861, Governor Curtain reconvened the legislature, issued a call for twenty-five thousand troops to quell the rebellion, and floated a $3 million loan to cover the cost of transporting and equipping the new soldiers. Democrats immediately brought charges of corruption and mismanagement to the public's attention, which forced Curtain to set up a commission to investigate the practices of the commonwealth's purchasing agents and the fairness of its supply contracts. The cost of arming, feeding, and transporting Pennsylvania's soldiers hovered somewhere between federal and state responsibility, but in the end the commonwealth was left with much of the debt. Over the course of the war, Pennsylvania spent over $5 million on the state's military units and received only $1.5 million in federal reimbursements. In fact, when the state floated a $3.1 million bond issue in 1861 to raise troops, the federal government refused to help at all.[8]

Pennsylvania's tumultuous political landscape of the 1860s combined with a boom mentality among colliers to reshape the wartime coal trade. The fuel needs of the federal army and navy, along with their military suppliers, promised a significant increase in the demand for coal. Mine operators planned for rising, or at least stable, coal prices for the duration of the conflict. Their expectations proved accurate. Even when prices are adjusted for wartime inflation, as table 6.1 demonstrates, they increased substantially over the course of the conflict. Over the years 1860 to 1863 the real price of a ton of anthracite rose by over 30 percent, and in 1864 the price had increased to 45 percent above its 1860 level. Wartime conditions favored the expansion of Pennsylvania's coal trade.[9]

Pennsylvania manufacturers appeared to weather the initial shock of shortages in labor and capital and spent the latter years of the conflict in steady recovery and growth. The state's coal trade, boosted by stable or rising prices for the first time in its brief history, followed the general pattern of Pennsylvania's manufacturers. By 1863 and 1864 rising expectations helped colliers overcome significant labor shortages and expand production. Pennsylvania colliers in both the anthracite and bituminous regions raised more tonnage throughout the 1860s. New firms accounted for much of this increase. But

Table 6.1. *Average Price of Schuylkill White Ash Lump Coal at Philadelphia, 1860–1865*

Year	Price per Ton (in dollars)	Price Index (1860 = 100)	Adjusted Price (in dollars)
1860	3.40	100	3.40
1861	3.39	103	3.29
1862	4.14	114	3.63
1863	6.06	136	4.46
1864	8.39	170	4.94
1865	7.86	193	4.07

Source: Israel Morris Jr., *Coal Price Current, Being Comparative Prices of Anthracite with American and Imported Bituminous Coal* (Philadelphia: F. Scofield, 1870); Stephen J. DeCanio and Joel Mokyr, "Inflation and Wage Lag during the American Civil War," *Explorations in Economic History* 14 (1977): 315.

what form would new coal mining firms assume? Would Pennsylvania's blend of special and general chartering of coal companies hold up under the unique political and economic conditions of wartime? Pennsylvania colliers encountered a legislature distracted by the financial and organization demands of war, wracked by partisan confusion, and reeling from the fallout of the tonnage tax. As a result, coal mining charters reached all-time highs in Pennsylvania during the Civil War.[10]

Without effective party discipline to regulate legislative activity, special incorporation in Harrisburg shot up to unprecedented levels by the later years of the war and threatened to topple the antebellum chartering regime that encouraged corporate development in some areas while protecting others from competition. In the 1864 session the legislature passed over 900 separate laws, and in 1865 it passed some 850 laws. The previous high for the Pennsylvania legislature was 732 laws in the 1857 session, thus the war years represented a significant increase in legislative activity. The Civil War years opened the floodgates for private legislation in Pennsylvania, as over a thousand laws were passed in each of the sessions from 1866 to 1872. The crisis mentality surrounding Harrisburg allowed the number of special charters passing the legislature to increase substantially as the war continued. Legislative activity in coal mining charters—measured in both new charters and charter amendments issued by legislative act—reflected the coal trade's optimism in the latter years of the war. In 1864 alone, forty-four mining charters passed the Pennsylvania legislature. Existing corporations also sought charter supplements that allowed them to purchase more land, issue bonds, or raise their maximum capitalization. "Special legislation is still the order of the day," reported one newspaper. "An act authorizing a company to operate with a million dollars capital goes through as readily as would an act to change the name of John Doe to Richard Roe. No remark is made—nobody votes against the bill and no person for it, save the Senator having the bill in charge."[11]

Such a free-for-all in charters and supplements created a system functionally similar to the mandatory general incorporation in place in New York and other northern industrial states, but a great deal of corruption still accompanied the chartering process in Pennsylvania. This made it difficult for colliers to predict which charters would successfully negotiate the legislature and which would not. A powerful illustration is found in the correspondence

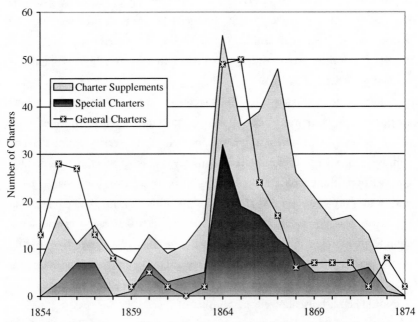

Fig. 6.1. Special and general incorporations in the Pennsylvania coal industry, 1854–1874. (*Laws of Pennsylvania, 1854–1874;* Pennsylvania Corporations Bureau, Letters Patent, 1854–74, Records of the Department of State, Pennsylvania State Archives, Harrisburg, Pa.)
Note: Supplements added to demonstrate chartering activity in the legislature. General charters include 1849, 1854, and 1863 general incorporation acts.

between George Magee, president of the Fall Brook Coal Company of Tioga County, and his agent in Harrisburg, Charles Lyman, during the legislative session of 1862–63. Lyman immediately reported to his employer that Fall Brook Coal Company's rival, the Tioga Railroad Company, was contesting its charter amendment and had employed a former senator to lobby against his bill by telling members that Magee was anxious to enrich himself at the expense of the people of Pennsylvania. Lyman also described a local legislator as, "our friend Beck [who] is one of the most dangerous we have to contend with here . . . if possible, he will *pinch* you till you *shell out*." Beck had suggested to Lyman that the Tioga Company was planning to spend eighty to one hundred thousand dollars to defeat the bill and that it was prepared to double whatever amount the Fall Brook Coal Company was spending. The object of this tactic, according to Lyman, was to "induce all venial men in the House to hold off—so as to be paid for voting for the bill." Although Lyman claimed that no money would be spent to procure this aim, he noted a few days later: "I have ordered the whiskey for Beck, he seems to be much improved in his feelings towards us. I told him that you regarded him as one of your most steadfast friends—how truthful that is[,] you know best." In the same letter he revealed that, although he was confident of the bill's passage in a straight fight, "what ever is to be done will of course be done in secret—And will only be developed when the vote is taken."[12]

Despite the heightened activity of the wartime legislature, special charters or supplements to existing charters continued to be as expensive to wrest from the Pennsylvania legislature as they had been before the war. A corporate rival's lobby, a rookie legislator, or a pesky trade journal could easily turn legislative sentiment against a charter and bring attention to its special privileges. This system ensured that a great deal of political clout was needed to allow certain charters to slip through the legislature. Disclosure of a charter's privileges or negative publicity could still undermine a firm's legislation. In 1861, for example, John Ewen of the Pennsylvania Coal Company wrote that "the passage of our bill so quietly, thro the Senate I confess surprised me, in view of previous experience in reference to former bills."[13]

During the latter years of the war, when chartering activity skyrocketed, public exposure could still cause problems. In 1864 a supplement to the Pennsylvania Coal Company's charter provoked the editors of the *United States Railroad and Mining Register* to proclaim that "corporations sometimes ply blandishments on those of the members who, in some cases, wear a thin disguise over notorious weaknesses." In this fashion many charters or supplements lingered in committee, failed to pass a floor vote, or were amended to death. In 1864 Governor Curtain vetoed a supplement to the Lykens Valley Coal Company charter which would allow the company to increase its landholdings from three thousand to fifteen thousand acres. In doing so, he argued that the situation had gotten out of hand: "I am determined to approve no bill creating a new monopoly of the kind, or giving one already existing the right of holding a larger quantity of land than they are now authorized to acquire."[14]

As special chartering increased during the war years, the number of corporations created under general laws also grew considerably (see fig. 6.1). In addition to the 1849 and 1854 legislation, an entirely new general incorporation act for mining and manufacturing passed through the legislature in the 1863 session. This act, which eventually spawned more than twelve hundred corporations, placed no restrictions on the amount of land companies could hold, fixed capital stock within the range of five thousand to five hundred thousand dollars (raised to one million dollars a year later), and allowed firms to own and mine land outside the commonwealth's borders. For the first two years the act did not apply to the important anthracite coal counties of Northumberland, Luzerne, Columbia, Carbon, and Schuylkill; also excluded were two western bituminous counties, Armstrong and Westmoreland.[15]

But, despite its liberal provisions, coal mining firms constituted only about 7 percent of the 1,236 corporations authorized under the 1863 act. The incredible rash of charters for oil companies, some claiming to hold as little as five acres while still capitalizing themselves in the hundreds of thousands of dollars, suggested that the 1863 act spawned many a fallacious business venture. The west's need for general and special charters persisted, as over three-fourths of the coal mining firms chartered under the 1863 act held lands west of the anthracite region. In fact, the act was originally limited to Allegheny County but later extended statewide. In its original form the general incorporation act of 1863 was intended to encourage development in western Pennsylvania, and it served that purpose well.[16]

Some state officials later regretted the liberal provisions of the act, even though it enjoyed widespread support among development-minded legislators and passed both houses with ease. Eighteen months after its passage Governor Andrew Curtain announced, "I approved the act in question with great reluctance and subsequent reflection and observation have satisfied me of its mischievous character." Curtain had indeed unleashed a monumental act for the western coal trade, as firms using the generous terms of incorporation offered by the 1863 act played a major role in the corporation reorganization of Pennsylvania's bituminous field. Special charters still required political and financial capital, but the sheer number of bills introduced and passed in 1864 and 1865 overwhelmed the antebellum system that issued general charters in developed areas and limited special charters to Pennsylvania's undeveloped coal regions. The rash of charters swept away any remnants of the anticorporate rhetoric in the legislature. More important, the blizzard of charters during the Civil War presaged a completely different organizational structure for the postbellum coal trade in Pennsylvania, even if legislators stopped chartering altogether.[17]

Existing coal interests found Pennsylvania's chartering atmosphere troubling by the latter years of the war. "Speculators opened the door for the introduction of Coal Mining Companies into the County by secret legislation unknown to the people," reported the *Miners' Journal* in 1865, while the *United States Railroad and Mining Register* noted that the 1863 general incorporation bill was passed by "the most scandalous body that ever assembled at Harrisburg" and that "countless bills affecting your private interests were BOUGHT through both Houses, just as you would buy a house or a horse."[18] It seemed to many observers that the geographical distribution of special and general charters, so effective in minimizing conflict between existing coal interests, had broken down. In 1864 the *United States Railroad and Mining Register* likened the lack of state control over the Pennsylvania coal trade to the "sudden, frequent, and disastrous floods" of the Ohio River and suggested that "*Laissez faire* is the curse of America, and the shame of the Yankee character." That same year the Harrisburg correspondent for the *Miners' Journal* observed that when "times are flush, the rage for speculation knows no bounds" and that corporate charters passed through the legislature like a hot knife through butter. "Time alone will correct the evil," he wrote a few weeks later, "but the wreck it will leave behind will bring sorrow to many a household." These comments proved prophetic, as the increase in charters late in the war and in the immediate postbellum years facilitated a rapid and enduring corporate transformation of Pennsylvania's anthracite and bituminous coal regions.[19]

Schuylkill County, despite its position as the bastion of individual enterprise and anti-charter rhetoric, witnessed the most dramatic corporate transformation during the Civil War. Advocates of corporate development in the Schuylkill region seized the opportunity to reorganize the area's independent collieries into large corporate concerns. W. H. Sheafer, an early innovator in deep shaft mining, served as the general agent for the Schuylkill region's most powerful new concern, the Mammoth Vein Consolidated Coal Company. This company, chartered in 1864 under the new general incorporation law, operated seven collieries that a year later mined 177,485 tons of coal. George Potts, who had once been one of the Schuylkill region's largest and wealthiest independent operators, also diverted his interests to corporate concerns. In 1865 Potts sat on the board of directors of both the Wolf Creek Diamond Coal Company and the New Boston Coal Company. George W. Snyder, a longtime independent operator whose Pine Forest colliery produced 39,978 tons of anthracite in 1865, assisted Potts with the New Boston Coal Company's mines. Snyder also helped organize the Locust Dale Coal Company in 1864, a firm that mined over 74,327 tons of coal the following year.[20]

Overall production in the Schuylkill region late in the war period confirmed the impact of corporate reorganization. In 1863 a single corporation, accounting for less than 1 percent the Schuylkill Region's tonnage, was listed among the mine operators in the *Miners' Journal*'s annual compilation of production statistics. By 1864 twenty-five corporations accounted for 22 percent of the region's operators and fully one-third of its production. Corporate mining accelerated in 1865, as fifty-two corporations accounted for nearly half of the region's operators and tonnage for that year. The corporate reorganization of the area's largest collieries occurred for the most part with charters issued in 1864 and 1865, as nearly half of the Schuylkill and Mahanoy regions' largest firms did not exist in 1863.[21]

The corporate form of organization also became prevalent in Pennsylvania's western bituminous fields. As mentioned earlier, western bituminous firms in the antebellum era often received "developmental" charters with extensive land grants and railroad privileges attached. During the Civil War corporate mining concerns continued to open new coal seams and accounted for a great deal of the increased production of western Pennsylvania's bituminous fields. The passage of the 1863 act, moreover, led to the widespread use of general charters in the western bituminous region. In the Phillipsburg semi-bituminous coalfield northeast of Pittsburgh, for example, corporate mining took off after the Pennsylvania Railroad integrated the region into its system in 1862. Annual production there increased from 8,000 tons in 1862 to 62,000 tons in 1865 and rose to 413,000 tons by 1870. Six corporations created between 1861 and 1867, two formed by special charters and two each formed by the 1854 and 1863 general chartering acts, accounted for 46.8 percent of the Philllipsburg region's production by 1871. The Pennsylvania Railroad facilitated the exploitation of the Phillipsburg region by companies created during the intense chartering activity triggered by the Civil War.[22]

Corporate ventures created in the 1860s increased production substantially in the semi-bituminous field of north-central Pennsylvania. Two Bradford County corporations created by the 1863 general act, the Towanda Coal Company (1865) and the Fall Creek Bituminous Coal Company (1865), accounted for all of the 393,023 tons of semi-bituminous coal shipped on the Barclay Railroad in 1871. In the same year another firm with a general charter from the 1863 legislation, the Morris Run Company (1864), accounted for 47 percent of the Tioga Railroad's coal traffic in Tioga County. The Barclay Railroad linked the Phillipsburg field to markets in central New York in 1856, and the Tioga Railroad completed its connection with the New York and Erie Railroad's system in 1853. Neither railroad saw a substantial increase in coal traffic until the Civil War, which suggests that the upswing in corporate mining during the mid-1860s greatly affected the development of this region.[23]

In addition to opening new fields, corporate mining also reconfigured existing bituminous coal regions during the Civil War. The liberal chartering regime during the war provided colliers in more developed western bituminous fields outside of Pittsburgh with the opportunity to apply corporate or-

ganization to their mining operations. For years colliers along the Monongahela and Youghiogheny river valleys used individual proprietorships to mine
coal. An 1859 directory of this bituminous region reported eighty-nine proprietorships and partnerships and one corporation, the Pittsburgh and Youghiogheny Coal Company, mining coal in the Monongahela and Youghiogheny
valleys. By 1871 the region had twenty-five corporate mining ventures and
fifty-seven proprietorships or partnerships—at least half of the corporations
received charters during the Civil War years. These corporations, even in areas still dominated by noncorporate firms, used the deep pockets afforded by
corporate capitalization to expand mining operations, purchase more land,
and integrate carrying facilities. Not surprisingly, these corporate firms soon
led the region in production. Along the Monongahela River's slack water navigation, twenty-four proprietorships or partnerships mined coal, as opposed
to one mining corporation, but that single corporation, chartered in 1865 under the 1854 general act, mined 100,000 tons of coal in 1871 and was by far the
area's largest producer.[24]

The corporate reorganization of Pennsylvania's coalfields threatened to
roll back the long-standing separation between mining and carrying privileges in coal corporations. The controversy between the Schuylkill Navigation
Company and the Lehigh Coal and Navigation Company had established the
ability to both mine coal and transport it to market as distinctly separate privileges in antebellum coal company charters. Such a division was necessary,
anticorporate interests argued, in order to preserve individual opportunity
in coal regions and prohibit price setting by monopolistic companies. A few
firms, such as the Delaware and Hudson Canal, and the Delaware, Lackawanna, and Western Railroad enjoyed both mining and carrying privileges
during the antebellum period, but newer coal carriers such as the Philadelphia and Reading and the Pennsylvania Railroad did not hold such provisions
in their charters.

Not satisfied with just transporting coal during the war, many railroad
companies appealed to the legislature during the sessions of 1863 and 1864 for
the right to buy or lease coal lands. Railroads had profited immensely from
the increased traffic spurred by the war effort. The Philadelphia and Reading
line, for example, saw its margin of profit increase from $0.88 per ton of coal

in 1861 to $1.72 per ton in 1865. In 1864 the Reading's stock price rose to 165 per-
cent of par value in New York, and the company declared a 15 percent dividend the same year. An 1861 law allowing railroad corporations to purchase
the stock of other firms loosened the chains a bit, but the legality of combin-
ing of mining and carrying rights remained unclear until 1869, when the
Pennsylvania legislature authorized the purchase of stocks and bonds of mining firms by railroad and canal companies. Governor Curtain's veto of two
charter supplements granting railroad privileges to mining firms in the 1866
legislative session suggests that this blend of carrying and mining privileges
had not yet become standard policy. But conventional wisdom regarding the
separation of these functions underwent a major revision, as the wartime
economy privileged increased production and efficient transportation over
the preservation of individual proprietorships in the coal regions. Railroads
serving the coal regions adopted an expansionistic strategy in the years immediately after the Civil War to exert a far greater direct influence upon mining firms than they had in the 1850s.[25]

Profits earned by railroads allowed them to consider expanding into mining coal; they also encouraged state officials to impose taxes on anthracite production. In early 1864 Governor Curtain singled out canals and railroads with
both carrying and mining privileges as "substantial monopolies" that make
"heavy gains at the expense of individuals." "In my opinion such privileges
ought never have been granted," he argued, "but as they exist, it appears to be
just that the class of companies which enjoy them should therefore pay an additional specific tax." In the special session of 1864, consequently, the legislature passed the commonwealth's first comprehensive state tax on coal. The tax
included all railroad, steamboat, canal, and slack water transportation companies in the state and imposed a rate of two cents per ton for "the products
of mines." It also taxed mining corporations not engaged in transportation 3
percent of their net annual earnings. Coal carriers readily, if not enthusiasti-
cally, accepted the new tax, at least as long as profits remained high during the
war. The major coal railroads and canals now constituted a critical source of
income for the Commonwealth of Pennsylvania. In 1865 income from the new
tonnage tax of $389,000 made up 6 percent of its revenues.[26]

The upswing in chartering during the Civil War boosted tax revenues for

Pennsylvania's state government and altered the commonwealth's revenue structure permanently. In antebellum decades the vast majority of tax revenue flowing into state coffers came from the slight taxes and personal and real estate. Most business corporations in Pennsylvania paid an initial tax on capitalization and continued to owe a very small tax on their dividends, depending upon their individual charter or the general laws under which they were organized. In the first full year of the Civil War corporate taxes made up roughly 8 percent of the state's revenues. By 1864, when the rampant chartering started to reorganize Pennsylvania's coal industry, corporate taxes made up roughly 13.5 percent of the commonwealth's revenue. The share of corporate taxes rose to nearly 20 percent of revenues, and in 1866 the roughly $1.26 million raised in corporate taxes eclipsed the amount gathered from personal and real estate taxes for the first time in history. Also in 1866, the combined contribution of corporate and coal tonnage taxes reached 28.5 percent of Pennsylvania's income. Over the relatively short time period of the late Civil War years, corporate tax revenues grew to assume a larger proportion of the commonwealth's revenues. This shift in tax revenues would have a tremendous impact upon Pennsylvania's postbellum political economy of coal.[27]

The growing power of organized labor in the anthracite fields redefined Pennsylvania's political economy of coal during the Civil War, much as the increase in chartering reorganized business structures and tax revenues in the coal trade. A new and more militant labor movement threatened the power of mine operators during the war. Throughout the early history of the anthracite region, miners employed about six to eight hands at most. Early anthracite miners also worked close to the surface, often in horizontal drift mines, which meant that work was not as dangerous in the era before deep-shaft mining. Most mining operations were far-flung enterprises away from urban centers, which frustrated attempts to organize miners into a "critical mass" of collective power. These factors, coupled with the mine operators' belief that individual enterprise in the anthracite regions ensured a competitive system of independent producers, had thus far inhibited the development of strong labor organizations in Pennsylvania's anthracite fields.[28]

Most disputes in the anthracite fields prior to the Civil War, when they did occur, were temporary affairs that focused upon the low wages spurred by the cutthroat competition in the anthracite trade. The first such action occurred in July 1842, when workers from Minersville in Schuylkill County marched on Pottsville to protest low wages. The Orwigsburgh Blues, a local militia company, broke up this short-lived strike. In 1848 John Bates enrolled five thousand miners and struck for higher pay in the summer of 1849. But members of the "Bates Union" found themselves locked out of work, and the movement quickly dissipated. In 1853 the Delaware and Hudson Canal Company's miners struck for a 2.5 cents per ton increase in their piece rate. This strike was successful, but it failed to produce any lasting union presence in the D&H's operations.[29]

On the eve of the Civil War labor relations in the anthracite region rapidly shifted as a number of trends converged to make working-class militancy in areas such as the Schuylkill region possible. First, mining operations grew in size, and coal regions became more populated. In 1842 there were between 3,500 and 4,000 miners working in Schuylkill County. This number had increased to about 10,000 by the mid-1850s and reached nearly 17,000 by the end of the Civil War. Second, the number of miners working for the same employer increased as a consequence of the corporate consolidation of coalfields during the 1860s. Of the 1,590 miners employed in Schuylkill County's Cass Township during the war, three-quarters worked in collieries owned by Charles Heckscher. Third, at this time the miners in Schuylkill County were much more likely to be Irish and propertyless than antebellum miners. Of the 1,811 Irish miners in Schuylkill County, only 153 (8%) owned property. Finally, the wartime demand for coal and relative scarcity of labor placed miners in an advantageous position to strike for higher wages. Because miners were among the lowest-paid industrial workers on the eve of the war, this potential loomed large in the minds of many observers.[30]

Working conditions for mine workers in the anthracite region worsened during the late antebellum years. A small-scale mine operated by an individual collier and a handful of paid laborers most likely operated close to the surface. As the easily excavated drift and slope mines gave way to deeper shafts that reached veins below the waterline, mine workers found themselves in vastly different surroundings. Water dripped constantly from the walls of many mine shafts, turning the floor of the mine into a slurry mess. Deeper shafts also required extensive ventilation systems, or else miners risked suffocation. Ventilation furnaces designed to pull foul air from the mines and draw fresh air from the surface helped somewhat, but the use of mules to tow mine carts to and from the face of the mine, combined with the dust and grime, lent a pungent quality to the air. Rather than lift them up and down the mine shaft every day, miners fed and sheltered their mules in the mine. Along with the mules came rats. Their constant presence at least indicated to miners that the fetid air would not kill them.

Miserable working conditions were not the only risks involved in anthracite mining. Anthracite mines adopted the breast and pillar system from the British mining industry but with some modern alterations. Skilled miners bored deep holes into the coal, filled them with black powder, and ignited them after shouting the standard warning, "Fire in the hole!" Explosions loosened the coal and made it easy to extract; they could also cause cave-ins or release deadly methane gas into the shaft. Once the main pockets of coal were removed, miners "won" the rest of the coal by extracting the pillars and replacing them with timber. Sometimes "timbering" could not adequately support the ceiling. If a miner spotted rats either running or swimming in a panic, he was wise to follow them away from potential explosions, cave-ins, and methane poisoning. Along with massive capital requirements that demanded corporate organization, deep-shaft mining in Pennsylvania's anthracite region exacted a high toll in life and limb.[31]

In order to improve wages and working conditions, many anthracite miners formed miners' benevolent associations during the 1860s. These local unions paid sickness and death benefits from monthly dues collected from their members, sent delegates to regional meetings, and raised money to support striking miners. They provided an organizational voice for miners not only to demand higher wages but also, in the words of historian Grace Palladino, "to shape an alternative model for industry, in which the rights and prerogatives of labor and capital had equal weight." Strikes broke out across Pennsylvania's anthracite regions in 1862 and 1863. In the face of high demand and labor scarcity, mine operators initially found it easier to meet the demands of their striking employees than to fight them.[32]

Pennsylvania's efforts to raise troops for the Civil War altered the balance of power in the state's anthracite fields. The violent resistance to the military draft created by the Conscription Act of 1863 provided a means by which mine operators could restore their brand of order to the anthracite fields and quell the rising power of the miners at the same time. A number of high-profile murders related to the widespread resistance to conscription among Irish miners and laborers provided mine operators with a means to impose their version of "law and order" in the anthracite regions. Federal provost marshals, operating under the federal prerogative to enforce the draft, called for federal troops to occupy Pennsylvania's anthracite regions and impose martial law. Ostensibly used to suppress draft rioters, these federal troops broke the nascent force of the miners' associations and allowed mine operators to gain the upper hand in the coal trade's newfound class conflict for the duration of wartime. Although the federal government played the prominent role in suppressing labor militancy in the coal industry, class conflict emerged as a critical component of Pennsylvania's state-level political economy of coal.[33]

Public officials in the Keystone State rarely intervened in industrial relations during the antebellum period, and their willingness to intercede on behalf of workers waned during the Civil War. In 1849 Pennsylvania passed the first in a series of laws that defined miners' wages as the first debts to be paid in case of a mine operator's bankruptcy, thus placing a "worker's lien" on the mining venture's land, lease, improvements, and property. The lien signaled a willingness to provide some relief for miners in the volatile coal trade, but attempts to draw the legislature into the causes of mine safety, hospitals for injured miners, and the injustices of the company store system all failed during the 1850s. A pattern evolved in which popular agitation and a barrage of petitions triggered the introduction of a number of rectifying bills, but mining interests killed them outright or in committee. Simply put, Pennsylvania's public officials appeared sympathetic yet reluctant to take action to protect the health and rights of miners. A significant example occurred during the wartime campaign to raise wages and prohibit mining firms from charging exorbitant rates to workers at company-owned stores. In his annual message to the legislature in January 1863, Governor Curtin argued that "it would be most unwise for the State to interfere at all with the rate of wages," but at the same time denounced the company store system as one that was "most unwise and unjust" and in need of legislative correction. A bill to eliminate these practices subsequently passed both houses in January 1864, but Governor

Curtain vetoed it because the law was limited in scope and probably did not apply to incorporated employers. The presence of the Union Army grafted a kind of temporary stability within the anthracite industry's labor relations. Workplace issues such as mine ventilation and safety inspection would, however, form the crux of class conflict in the postwar coal trade.[34]

In sum the unique conditions of the Civil War—high demand, more charters, and a forcefully subdued workforce—triggered a number of permanent changes in Pennsylvania's political economy of coal. The boom in corporate charters and subsequent reorganization of the state's most ardently anti-charter anthracite field provided a definitive answer to Pennsylvania's long-standing debate over the future of corporations in the state's coal trade. The idea that mining was best left to individuals became a distant memory in post-bellum sessions of the Pennsylvania legislature. Wartime politics also introduced new controversies such as the presence of a nascent labor movement in the anthracite region, the ability of mining firms to control the workplace, and the dependence of state government upon corporate revenue taxes. Whether by design or accident, the Civil War years set Pennsylvania's coal trade on a different course. Postbellum politicians now faced a coal industry dominated by large railroads, corporate mining firms, and an increasingly militant labor force. The days of protecting proprietary colliers and promoting the expansion of the trade were gone; in its place a political economy based upon mediation between government, capital, and labor would emerge in the Keystone State.

COAL AND THE CONFEDERACY

In eastern Virginia the Richmond basin briefly proved to be an important mineral-producing region, but the fuel demands of the Confederate war effort only temporarily revived the region's economic prospects. Richmond basin production hovered around 100,000 tons annually on the eve of the war, with traffic divided among the James River and Kanawha Canal and three local railroads, the Clover Hill Railroad, the Richmond and Petersburg Railroad, and the Richmond and Danville Railroad. Pennsylvania anthracite's domination of urban markets along the eastern seaboard by this time limited Richmond coal's markets to local smiths and small manufacturers. The Tredegar Iron Works, the area's largest manufacturing firm, imported 1,500 tons of anthracite for use in its furnaces and foundries in 1859. By the next year only two firms, the Midlothian Coal Company and the Clover Hill Coal Company, mined coal on a large scale in the Richmond basin.[35]

The outbreak of war, however, cut off Virginia's supply of anthracite and therefore revived the strategic importance of the Richmond basin, at least temporarily. The Tredegar Iron Works, now the major ordnance supplier for the Confederate war effort, needed new sources of mineral fuel. Joseph Reid Anderson, the senior partner at the Tredegar Works, secured an adequate supply of bituminous coal from the Midlothian and Clover Hill companies through December 1862 but needed more than existing mines could furnish. As the orders for more armaments rolled in, Anderson arranged for a $200,000 loan from the Confederate War and Navy departments to purchase and operate the Dover and Tuckahoe mines in the Richmond basin. In exchange for the loan Tredegar agreed to sell any excess coal to the government. The Tredegar Works suffered shortages of pig iron, food, and other essentials throughout the war but found Richmond basin mines an excellent source of coal.[36] Area mining firms also prospered. In 1863, for example, the Clover Hill Coal Company reported a gross income of $136,931.29 for the year, which exceeded its expense paid in capital of $100,000.[37]

In 1862 the Confederate Congress organized the Niter and Mining Bureau within the War Department to supervise the collection of niter (also known as saltpeter) for the manufacture of gunpowder and the mining of copper, lead, iron, coal, and zinc. In the Richmond basin private firms remained in control of their property, and the Niter and Mining Bureau served mostly to allocate desperately needed slave and free labor to mines across the South. In addition to aiding the Richmond field, the Niter and Mining Bureau opened new coalfields in North Carolina and Alabama and coordinated the flow of mineral fuel to Confederate naval stations along the coast.[38]

Despite the best efforts of state and federal officials, however, coal production outside of Richmond eventually fell prey to the disrupting presence of the Union Army. In March 1864 the manager of the Dover Coal Pits, Charles Quarles Tompkins, recorded in his diary a "mission of plunder, fire, and destruction" upon his works by an advance party of federal cavalry. The cavalry unit burned the woodwork on several shafts and confiscated seven slaves. Throughout the spring and summer of 1864 federal troops raided the area to secure horses, mules, and other supplies. The Union Army brought physical devastation to the area and encouraged many slaves to flee to federal lines. But, despite chronic labor shortages and the constant threat of invasion, Richmond basin colliers still shipped a significant amount of coal. Tompkins noted that the Dover Pits raised coal within earshot of heavy cannon fire, and in 1864 Richmond area miners raised 112,068 tons of bituminous coal. Relatively high production levels continued until the siege and subsequent evacuation of Richmond effectively ended wartime demand for local coal in April 1865.[39]

Plans to revitalize the Richmond basin following the war, however, fizzled. Joseph Reid Anderson sold Tredegar's share of the Dover coal tracts to New England investors in 1866 but was unable to sell the company's share of the Tuckahoe pits, due to the "unsettled political condition of the country." Investment and development in the Richmond basin remained sporadic throughout the postbellum years. In the early 1870s Oswald Henrich, the superintending mining engineer of the Midlothian colliery, argued that the region's problems stemmed from "ignorance, want of system, and false economy," among eastern Virginia mining firms, which "concentrated the worst elements imaginable to prevent the continuous prosperity of the mines." Considering their proximity to canal and railway links, Richmond colliers appeared particularly unfortunate by the late nineteenth century. In fact, the area never adequately recovered from its early competition with Pennsylvania anthracite and, despite a brief resurgence in strategic importance during the Civil War, the Richmond basin never lived up to its potential. "An unaccountable lethargy seems to have fallen on all who touched this field," the *Engineering and Mining Journal* wrote of the Richmond basin in 1876, "and yet, in the future it can scarcely fail to become one of the great sources of coal supply."[40]

THE CREATION OF WEST VIRGINIA AND THE COAL TRADE

Although coal production in Virginia's western counties, like in the Richmond basin, stalled as a result of hostilities, the creation of West Virginia established an entirely new institutional context for the industrial development of western Virginia. Just as the Civil War facilitated the corporate transformation of

Pennsylvania's coal trade and demonstrated the remaining potential of Richmond's bituminous coal mines, West Virginia's creation occurred as the result of the war's military and political conditions. Nevertheless, the idea of separate statehood and its importance to the region's economic future had deep intellectual and political roots in the mountains of western Virginia and important implications for the shape of West Virginia's postbellum coal trade.

The creation of West Virginia is rooted in Virginia's fierce sectional rivalry. The experience of most states in the Upper South, particularly Maryland and Tennessee, suggest that the Old Dominion was not the only place where eastern and western interests clashed. But, unlike other states in the Upper South, Virginia's sectional struggle paved the way for the uncommon solution of dividing the Old Dominion into two states during the Civil War. The political frustration of western Virginians translated into calls for separate statehood more than thirty years before West Virginia entered the Union. The future of representation in Virginia—or at least the political power of well-propertied easterners in state government—served as the touchstone of these persistent demands for a separate polity. Controversies over internal improvement funding, political representation, and corporate chartering waxed and waned in the Virginia legislature, but the political influence and economic position of eastern slaveholders remained a constant, although often muted, basis of sectionalism in the Old Dominion.

Western Virginians, in fact, had little in common besides their opposition to Richmond's influence in their affairs and their interest in promoting western Virginia's mining and manufacturing economy. Westerners in the Kanawha Valley, although quick to criticize the aristocratic tendencies of tidewater politicians, had few problems with the institution of slavery as a labor system and very much considered themselves southerners. Virginians of the northern panhandle, with its intellectual and economic center in Wheeling, identified much more readily with the free labor ideology of their neighbors, Pennsylvania and Ohio. The Kanawha and Panhandle Virginians formed alliances in Virginia's legislature and constitutional conventions, but southern West Virginians still had a "large leaven of secession," in the opinion of Granville Hall. Although they had many cultural and political differences, western Virginians felt equally marginalized in Richmond's conservative polity. The political frustration of westerners evolved into a sectional interpretation of events that inevitably led to independence. So long as easterners held the preponderance of political power in the legislature, western interests would suffer. Given that the existing power structure appeared so immune to meaningful reform, this sectional theory posited, eastern interests could continue to block the development of mining and manufacturing in the west indefinitely. Despite the significant differences between northern and southern outlooks within western Virginia, a common political cause between these counties emerged over the antebellum years.[41]

The idea that separate statehood offered a solution to Virginia's sectional problem emerged as early as the state constitutional convention of 1829–30. As delegates in Richmond delivered eloquent speeches concerning the rights of man and property, some western Virginians seethed with resentment. One editorial published in Charleston's *Western Virginian* denounced the "terrapin policy" of eastern Virginia's "lowlanders." "I am not yet prepared to believe that the Great Jehovah has placed mines of wealth in our hands to curse us with—as he has nations of people—," an editorial writer named Salt Lick argued with a none too subtle allusion to eastern Virginia's slaves, "but if we continue to abuse our heritage, let the righteous flee from Soddom." In 1830 a series of editorials under the byline of Senex advocated "civil discord" in response to the lukewarm reforms enacted by the convention. The anonymous writer recommended nothing less than the "ultimate division of the State." "Possessing the raw materials for manufacturing, nature seems to have designed western Virginia for a prospering and happy people," Senex argued, "and, with a government, suited to their genius, nothing is more certain, than that this design of nature would be realized to an extent enjoyed by but few communities on the face of the earth."[42]

In 1847 Reverend Henry Ruffner, the president of Washington College in Lexington, eloquently presented a synthesis of separate statehood, antislavery, and racism in his widely discussed pamphlet *Address to the People of West Virginia*. Ruffner recounted the tension between eastern and western Virginians and argued that emancipation was the best solution. He compared the prosperity of free states in agriculture, commerce, and manufacturing to the "stagnation" and "positive decay" of the slave states, where one found, "instead of the stir and bustle of industry, a dull and dreamy stillness, broken, if broken at all, only by the wordy brawl of politics." Ruffner reasoned that slavery innately promoted indolence and inactivity among whites and thus hampered the development of both agriculture and manufacturing. Furthermore, because the state's slave population naturally increased over time, the price of slaves would eventually fall. Once this happened, the reverend predicted, Virginia's investment in slave labor would be worthless and its black population a burden to white citizens. The termination of slavery in western Virginia, Ruffner concluded, would trigger an influx of capital and people to "draw out the wealth of its mines, and make the idle waterfalls and coal beds work up its abundant materials of manufacture." Ruffner's pamphlet was condemned as an abolitionist document by the newspapers of eastern Virginia, even though he advocated a conservative plan of gradual emancipation and colonization of freed African Americans to Liberia. Whether western Virginians agreed with emancipation or not, the pamphlet became a rallying point for those who regarded the political influence of eastern Virginia as the major impediment to the economic growth of their region.[43]

The controversy over tax policy in late antebellum Virginia illustrates this point. The inequality of the taxation of slavery—regarded as such because the state taxed slaves more lightly than other forms of property—emerged as a major focus of western Virginia's complaints after the "Reform" Constitution of 1850 reconciled many western concerns over suffrage and provided for a more democratic apportionment in Virginia's House of Delegates. Once again, politicians from western Virginia decried slavery's influence, this time in reference to the financial burdens of taxation. "I am a slave holder, and I regard the title to this property, as to all other property, as sacred, equally sacred," Waitman Willey of Monongalia County announced to slaveholding Virginians in 1851. "But I cannot allow my interest, in this respect to override the natural rights and liberties of one hundred thousand of my fellow citizens. No! You, a mere minority of this people, claim authority to legislate for the majority—to control their interests, civil, religious, political, even life itself. You have no right to such control." The system of exempting slaves under the age of twelve from taxation and valuing those over that age as equal to the tax on $300 in land revived Ruffner's criticism of the political economy of slavery in western Virginia. In the northwestern panhandle complaints regarding the inequality of Virginia's tax system seemed especially bitter. "It is a terrible sin, says the slavery propagandist, to say aught against the institution of slavery, or the abuses of the system," argued the *Wheeling Intelligencier* in 1860, "but it is perfectly right for those same slavery propagandists to tax, without reason and without mercy, the co-heas [Cohees] or poor devils of Western Virginia, for the especial benefit of their favorite institution." Virginia's unequal tax structure became a symbol of state-level sectionalism's influence in the Old Dominion in national, and even international, antislavery journals. But, most important, it kept ideas of separate statehood salient for western Virginians throughout the antebellum years.[44]

The negative reaction of eastern Virginians to the election of Lincoln in 1860 and the subsequent secession of South Carolina drove an even stronger wedge between eastern and western Virginians. In February 1861 a statewide convention met in Richmond to decide whether the Old Dominion would secede and join the Confederacy. As an enticement for delegates from western Virginia, organizers of the secession convention promised that tax reform would be included in the discussion. An early vote for secession on 4 April failed by a forty-five to eighty-five count, but Lincoln's call for 75,000 troops and the attack on Fort Sumter eroded support for the pro-Union position. On 15 April former governor Henry Wise, wild-eyed and brandishing a pistol, announced that Virginia troops were seizing federal property as he spoke. The next day the convention voted eighty-eight to fifty-five to adopt an ordinance of secession, which was ultimately ratified by the voters of Virginia during the spring elections in May 1861.[45] Of the forty-seven delegates from counties that later formed West Virginia, twenty-eight voted against secession, fifteen voted for secession, and four abstained from voting. Western delegates to Virginia's Secession Convention such as Waitman Willey and Thomas Carlile were among the leaders in the move to keep the Old Dominion loyal, and their votes reflected the desire of most western Virginians to remain in the Union.[46]

Unionist politicians, unwilling to accept secession, gathered at Wheeling in the summer of 1861 to form the "Reorganized Government of Virginia," which denounced the legitimacy of secession, declared loyalty to the Union, and seated members in both the Federal House of Representatives and the

Senate, the latter being Waitman Willey and Thomas Carlile. In August 1861 the Reorganized Government passed "An Ordinance to Provide for the Formation of a New State Out of a Portion of the Territory of this State," which provided the authority for Congress to recognize West Virginia two years later. But only the military invasion of western Virginia, which began in May 1861, made separate statehood a viable option for Unionists in the region. After General George McClellan's brief campaign to secure the area under Union control, western Virginia remained a relatively unimportant military theater of the Civil War. The organization of irregular troops, commonly known as "bushwhackers," however, resulted in a series of unsettling raids on the commercial and transportation centers of the area. The most famous, the Jones-Imboden Raid of 1863, resulted in the demolition of two railway trains and sixteen railroad bridges of the B&O line, the capture of 1,000 head of cattle and 1,200 horses, and the destruction of equipment and 150,000 barrels of oil in the Burning Springs field. Violence and confusion continued to plague West Virginia through the end of the war, as local antagonists used the Union or Confederate banner to settle old scores and create new ones.[47]

Amid the confusion of secession and war, fifty-three delegates from thirty-nine counties in western Virginia met in Wheeling in November 1861 to draft a constitution for the new state. West Virginia's founding generation came from a volatile mix of farmers, religious reformers, and lawyers. The majority of delegates from southern counties of western Virginia sought to replace Richmond's government with a new regime made up of similar institutions. As the long history of Virginia sectionalism demonstrated, western problems with the Old Dominion revolved around representation and power, not the structure of government itself. But the convention also boasted a sizable block of ministers, led by the abolitionist Reverend Gordon Battelle of Clarksburg, who attempted to reshape the political institutions of western Virginia so completely that the new state would resemble its northern neighbors. These evangelical delegates provided the moral leadership for the political faction known as the Radicals, which would dominate West Virginia's postwar Republican Party. Delegates from across the state concurred that the development of mineral resources of the new state required prompt attention, but the similarities ended there. A number of critical elements of West Virginia's political economy of coal bore the imprint of the convention's blend of religious reform, economic boosterism, and the Virginia political tradition.[48]

The coal trade, or at least its potential, played a large role in dictating the very boundaries of the state, as seen in the inclusion of counties along the route of the Baltimore and Ohio Railroad in West Virginia. The original constitutional convention drew delegates from thirty-nine counties west of the Allegheny Mountains—the traditional definition of western Virginia. The Committee on Boundary recommended the addition of thirty-one counties, including some that were west of the Alleghenies and others that were not.[49] Among the most controversial of those east of the Alleghenies were Morgan, Berkeley, and Jefferson counties, which today make up West Virginia's eastern panhandle and through which the Baltimore and Ohio Railroad ran. Antislavery delegates opposed their inclusion on the grounds that they would substantially increase the number of slaves in the new state. Defenders of the proposal, such as Waitman Willey, pointed out that the future of the state depended upon the B&O Railroad. "It conveys into our center," Willey argued, "or by its ramifications of necessity infuses through the entire body politic of this new State the life-blood of its existence." If they left the B&O to run through the Old Dominion, they argued, the Virginia legislature would punish both the railroad and the new state. "If we do expect to derive prosperity from this separation; if we do expect by being allowed to form our own institutions and conduct our own business in our own way, and attain that degree of prosperity which all hope for," argued delegate Peter Van Winkle of Parkersburg, "certainly it would be but a suicidal policy to throw away, cast from us, the very instrument by which all this good is to be effected." Moreover, one delegate from Marion County pointed out that without the B&O the state's minerals would "lie there as worthless as a common rock." Considering the coal trade, he argued simply, "the destruction of that road is the destruction of that interest." These pragmatic appeals, coupled with the assurances that slavery would not be a part of West Virginia's future, won over the votes necessary to include the eastern panhandle in the new state of West Virginia. At the basic level of state formation, the creation of boundaries, a sizable portion of delegates saw the future of the B&O Railroad as inextricable with those of West Virginia.[50]

The second major issue of political economy at the convention was the construction of a policy regime for the creation of corporations. A proposal to force the mandatory public disclosure of special charter applications in the West Virginia legislature provoked a significant debate over the future of charters in West Virginia. James Henry Brown, a lawyer from Charleston, found corporations "essential as the sun or the changing seasons almost to our national existence, to our institutions—a part and parcel of our society," which occurred for "the benefit of the community" and were therefore undeserving of such an unnecessary burden. Delegate Chapman Johnson Stuart of Doddridge County in the northwest noted the "influence and control the incorporated companies have had over our state legislation" and argued in favor of both the restriction and general charters. "Why these incorporated companies, sir, have been almost the ruin of our country," Stuart barked. "Give the people notice and stop this legislation, this kind of 'log-rolling' which is gotten up at the end of a session. I never want to see such a state of affairs inaugurated in the new State of West Virginia." Over the course of the debate delegates invoked the manufacturing laws of states such as Pennsylvania as potential models for West Virginia, but they also took care to notice that few states had completely eliminated special charters. A system modeled on Pennsylvania's blend of general and special charters won out in the end, as the convention made general charters mandatory for mining and manufacturing operations but allowed special charters—with the public disclosure of their application in the legislature made mandatory—in the case of internal improvements. The apparent consensus in favor of corporate development emerged as a critical component of the state's future political economy of coal. The necessity of corporate mining in West Virginia stood unchallenged during the state's early years of existence.[51]

A more divisive discussion erupted over the taxation of corporations in West Virginia. Both conservative and reform delegates concurred that corporations would play a large role in developing the state's natural resources, but they disagreed over the method of taxing these institutions. Reform delegates under the leadership of Gordon Batelle advocated using corporate tax revenues to construct a free public school system in the fashion of Ohio or Massachusetts. But the debate quickly turned from a discussion of the destination of revenues to one concerning the extent of revenues to be derived from corporate taxes. Benjamin Smith of Kanawha County, who had sponsored Virginia's circuit court chartering legislation nearly a decade earlier, was quick to attack a proposal that created corporate taxes to fund West Virginia's school system. Because the young state was short on capital, he argued, "we ought therefore so to direct our legislation and our action as to arouse people to call forth all the energy and power of the State to develop its resources." "I do not like to see measures taken to cripple corporations," Smith continued, "which are nothing more nor less than the associated wealth of the people of the country, who unite together their moderate resources to make a capital that will produce a good result." Without incentives to attract capital such as low corporate taxes, like-minded delegates argued, West Virginia would remain forever poor and undeveloped.[52]

Reformers agreed that corporations were indispensable to West Virginia's economic future but argued that corporate privileges bestowed by the legislature should come with responsibilities to the people of West Virginia. "There is such a thing, sir," William Stevenson of Parkersburg argued, "as corporations which may be of great public advantage under some circumstances becoming a great public nuisance under other circumstances." Over the course of the convention, conservatives defeated the proposition to fund free schools through corporate taxation by evoking nostalgic images of locally controlled schools and reiterating the idea that corporations simply would not invest in a tax-heavy West Virginia. The constitution eventually provided for the state support of schools through legislative appropriation and the sale of delinquent lands. The debate over educational taxes reinforced the dominant opinion that corporate investment—whether in the form of existing firms such as the B&O Railroad or in future mining firms reluctant to pay state taxes—would be sacrosanct in the new state.[53]

The pledge of West Virginia's state credit to internal improvement programs constituted another controversial element of state formation in the 1861 convention. This debate, like many others, revealed the different interests of delegates from the northwestern portion of the state and those from the Kanawha Valley. The Virginia Board of Public Works' politically motivated funding of internal improvement programs formed an enduring critique of the ancien régime, but convention delegates from northwestern Virginia had little reason to seek more state funding for railroad connections. Wheeling received its link to eastern cities in 1853, when the B&O completed its main line to the Ohio River. Four years later the Northwestern Virginia Railroad (a subsidiary of the B&O) connected Parkersburg on the Ohio River to the B&O's main line at Grafton. Delegates serviced by the B&O system saw little need for the new state to be burdened with a massive public works program. After all, they argued, the problems suffered under the Virginia Board of Public Works seemed inherent to state-supported ventures, so why repeat the same mistakes all over again? Ohio and Pennsylvania arose as examples of states that "have attested to the folly of this system of internal improvement on state account" and of good models for the new state. "Make a good constitution, under which capital will feel safe," exhorted a delegate from Parkersburg, "under which capital and enterprise will be encouraged and protected, omitting all this wild folly of trying to use the credit of the State to build railroads into country that can offer them no business until it has been created."[54]

Delegates from the Kanawha Valley had very different interests. In 1860 Virginia's General Assembly had increased the capitalization of the James River and Kanawha Company to $12.4 million and allocated $300,000 for the improvement of the Kanawha, but the war made this much-needed infusion of capital irrelevant. Delegates from southern counties therefore hoped that the new regime would assume their long-standing goal of improving the Kanawha River and its tributaries to accommodate the area's coal trade. Or, if the river improvement project still reeked of Richmond's logrolling and indecision, the state should at least help with a railroad connection. Benjamin Smith of Charleston spoke for many in West Virginia's southern counties. "We come up here in good faith to unite with you in forming this new State," he claimed, "but we never contemplated a connection with a people who would put in their Constitution, in their fundamental law a principle prohibiting that State from ever improving her own condition." The Kanawha position lost at first, as a provision for using state bonds to fund internal improvements was defeated by a strictly sectional vote of twenty-three to twenty-five. Many northwesterners feared, however, that southern delegates would forsake separate statehood entirely and pushed to reconsider the issue. Henry Dering of

Monongalia County helped hammer out a compromise provision that allowed the state to invest in internal improvement corporations so long as stock subscriptions were paid with cash or from taxes levied in the ensuing year.[55]

The convention remained relatively silent about slavery for most of its existence, even though critiques of the slave system had formed the basis for separate statehood in the first place. Robert Hagar and Gordon Battelle, both ministers in the Methodist Episcopal Church, introduced resolutions abolishing slavery, but both were laid on the table, and a tacit gag rule governed much of the convention. Under Battelle's persistent guidance the convention finally agreed to place a much watered-down emancipation proposal as a separate ballot item during the ratification vote.[56] Radical Republicans in the U.S. Congress and President Lincoln opposed West Virginia's foot dragging on slavery and eventually tied admission of the state to its abolition of slavery. Waitman Willey, then representing Restored Virginia in the Senate, attached an emancipation amendment to the new state's admission legislation which ultimately resulted in the acceptance of West Virginia to the United States in June 1863. Although West Virginians made clear their intention to rely upon out-of-state railroads such as the B&O to develop commerce and out-of-state capital to mine their coal, it appears puzzling and not a little foreboding that the state's relationship to slavery—one of the most significant unifying factors in its creation—was also imposed upon West Virginians from sources outside their boundaries.[57]

At any rate, wartime conditions ensured that the new state's coal trade would begin practically from scratch. Unlike their colleagues in the Richmond basin, colliers in the Kanawha Valley had a difficult time sustaining coal production in the face of war. The transition from formal lines of battle to the more informal "raid" in western Virginia after 1861 meant that military actions often targeted property and improvements more than enemy troops. Annual production in the Kanawha Valley dropped from 129,000 tons in 1860 to 88,000 tons in 1863 and bottomed out at 68,600 tons in 1865. Colliers along the Ohio River fared better, as annual production in Brooke County in 1865 doubled its 1860 level. A series of strikes centered at Wheeling along with a chronic labor shortage in the region, however, limited the expansion of western Virginia's Ohio River coal trade during the war. As a result, the expected boom in the northern panhandle never materialized. Overall coal production in the counties that became West Virginia dropped to under 300,000 tons in 1862 and 1863 and did not reach 1850s levels again until the final year of the Civil War.[58]

Radical West Virginia's state formation thus transpired over strange circumstances. The convention's tense resolution of the slavery question, as with the boundary issue, corporate chartering, and taxation, mirrored the debate over the state government's role in internal improvement and demonstrated that West Virginians shared different outlooks on the structure and character of their new state. In fact, the convention debates revealed a number of sectional positions that defied contemporary laissez-faire or regulatory positions. Kanawha Valley delegates supported active chartering and low corporate taxes, and yet they still expected the state to pledge its credit for internal improvements. Delegates from the northwest wanted to tax corporations to support public education but also openly championed the interests of the B&O Railroad. In 1863 western Virginians realized their desire to be rid of the slavery-addled politics of the Old Dominion but struggled in defining the shape their new state would assume.

In many ways the Civil War marked a radical departure in the political economy of coal in Virginia and Pennsylvania. A distracted and, at times, openly corrupt legislature stretched Pennsylvania's flexible policy-making structure to the limits. As a consequence, the state's coal trade rapidly changed from a blend of small entrepreneurial ventures and corporations to an industry dominated by corporate capital. Labor also emerged from the war years as a more powerful element in Pennsylvania's political economy of coal. Although temporarily held down by Union troops, militant unions challenged the coal operators' new corporate order in the years following Appomattox. The appearance of large mining and carrying companies upon Pennsylvania's political landscape also recast the Keystone State's policy-making structure. In the postwar years these large corporations would dominate Pennsylvania's political economy of coal, often with the complicity of the state legislature. The Civil War thus served as a watershed in which controversies that characterized its antebellum political economy of coal—transportation and mining privileges, for example, as well as corporate chartering—embarked on new directions for the postbellum decades.

The Civil War also ushered in massive changes in the Old Dominion's political economy of coal. Virginia's Richmond basin enjoyed a brief period of strategic importance during the war but continued its relative decline in the years afterward. The more dramatic change in the Old Dominion transpired west of the Alleghenies. Western Virginia, once the poor relative of the tidewater and piedmont, won political independence during the Civil War but struggled to develop its mineral wealth. Colliers in that region constantly agitated for a larger role in Virginia's economic policy making. Once they received a state government of their own, however, the dictates of the market and the lack of local capital made the rapid growth of a vibrant coal trade difficult to nourish in West Virginia. In the end the political and economic patterns established in antebellum western Virginia plagued the new state during its early existence.

Epilogue

Capture and Confusion

*I*n 1866 U.S. Senator Waitman Willey addressed an enthusiastic crowd of West Virginians in the railroad town of Fairmont. From his log cabin origins in rural Monongalia County to his pro-western stance in both the Constitutional Convention of 1850 and the Secession Convention of 1861, Willey spoke with the gravitas of a founding father to the young state of West Virginia. In the victorious afterglow of western Virginia's separation from the Old Dominion, policy makers hoped to reshape their state in the image of Ohio, Pennsylvania, or New York. In order to accomplish this feat, Willey proclaimed that the aim of all West Virginians should be the construction of a "capital-friendly" state. At that very moment political officials in Wheeling were promoting legislation that eased restrictions on out-of-state corporations and cozied up to large railroads such as the Baltimore and Ohio. "Let us properly invite capital from those who have it to spare," Willey announced to the assembled crowd. "If we take the necessary pains to inform capitalists of the actual natural wealth of our State they will, for their own sakes, come to our relief."[1]

Willey's speech reflected the ongoing belief that state government officials should mold their local institutions so as to maximize economic development. Although capital might originate from private sources, it was the responsibility of public institutions to direct it to their particular interests. Most states that had adopted this strategy in the antebellum years had withdrawn financial support for such activities by the end of the Civil War. The failure of state-sponsored internal improvement networks such as the Main Line and the concurrent rise of powerful railroad corporations such as the Philadelphia and Reading, the Baltimore and Ohio, and the Pennsylvania railroads recalibrated the parameters of state-level political economy. The shape of many Radical Reconstruction governments in the former Confederacy, moreover, triggered a dramatic shift in the regional patterns of state-level policies, as many southern states embarked upon ambitious growth strategies at the same time that the idea of "laissez-faire" policy making gripped northern statehouses. Industrial interests thus began to reshape completely antebellum notions of the political economy of coal in the postbellum decades—so much so that many scholars refer to the "capture" of public institutions by private interests during this period.[2]

The emerging role of the nation's federal structure in the decades following the Civil War provided new challenges for colliers in Pennsylvania, Virginia, and West Virginia. Preserving entrepreneurial opportunity, promoting the use of mineral fuel, and underwriting the rapid exploitation of new coalfields had defined capital-friendly policy in the 1830s or 1840s. By the 1860s and 1870s corporations dominated most coalfields, Americans burned coal at unprecedented rates, and new coal deposits awaiting discovery lay across the Mississippi River. New issues such as mine ventilation, labor relations, and corporate tax revenues dominated legislative agendas, while huge railroads, interstate mining corporations, and labor unions materialized as new actors on the policy-making stage. State-level activism did not disappear in this new environment; it became transformed into a new form of regulation which sought to preserve high production levels at a minimum cost to consumers.

The story of coal in "Radical" West Virginia illustrates this new context well. Armed with antebellum ideas about fostering the exploitation of coalfields, West Virginia's state officials learned quickly that encouraging internal improvements, geological surveys, and corporate chartering in the postwar economy was not a panacea. The models for antebellum development, such as Pennsylvania, could not meet the demands that postbellum policy makers faced. The demand for coal continued to climb in the postwar decades, but, as West Virginians quickly realized, the actors in charge of development were more likely to reside in New York, Boston, or Philadelphia than in Harrisburg, Richmond, or Charleston. The resulting coal trade did not benefit the fledgling state government; nor did it provide the expected benefits of tax revenues, increased commercial activity, or industrial development in close proximity to the state's coalfields. When he announced West Virginia as being capital friendly, Senator Willey misunderstood the scope and meaning of his own words.

"THE CHIEF STAPLE OF THE STATE"

Coal trade pundits responded favorably to the call for development issued by Willey and other prominent West Virginians, as the state's rich bituminous fields received immediate national attention. Philadelphia's *United States Mining and Railroad Register* announced that "several companies have been formed, and some of them got into operation" in 1865, and that bituminous coal, along with iron, should "constitute the chief staple of the State" now that the "lamentable political conditions" of the Old Dominion have been removed. Pottsville's *Miners' Journal* added that the Kanawha Valley, which had been "cursed and controlled by the slave masters who 'like the dog in the manger,' have for fifty years denied it to enterprise, and knew not how to profit by its immense mineral wealth themselves." Now, however, Pennsylvania's leading coal trade journal noted, the region was "open to free labor, capital, and enterprise." Once West Virginia's coalfields emerged from the slave system's heavy grip upon internal improvements, capital, and entrepreneurship, northern observers suggested, their development would rapidly progress.[3]

A brief glance at their counterparts in eastern Virginia might have tempered the enthusiasm of West Virginia's coal interests. Although slavery had disappeared, its lasting legacy affected colliers in the Richmond basin for decades after emancipation. Production declined to levels below 100,000 tons in 1869, and colliers in eastern Virginia deplored the weakened state of their trade by the 1870s. The potential wealth of the field appeared obvious. Job Atkins, an engineer hired to examine the Burfoot tract belonging to the Wickham family, claimed that the area "possesses advantages for successful mining operations that are enough to induce capitalists to buy it in preference to any other" and that its value "cannot be overlooked by the observant public." But even in the years following emancipation the Richmond field labored under the legacy of slavery. James M'Killop, a Scottish collier, visited the Richmond field in 1869 during a tour of American coal operations. He later described the valuable reserves found in eastern Virginia but posited that slave labor used in the mines prior to the Civil War had spoiled the Richmond field. "The system of its development places it behind that of any other country," M'Killop wrote, "since neither skill nor science had apparently aided brute force with any degree of intelligence."[4]

With the Richmond basin's growing irrelevance, Virginians looked to their mountainous southwestern region for mineral wealth. But the rich deposits of coal that underlay Tazewell and Wise counties owed their early development to individual developers and railroads rather than political institutions. Affairs in Richmond during the 1870s revolved around the political career of William Mahone, the leader of the "Readjuster" movement created to alleviate the Old Dominion's financial woes. Since Virginia defeated the most radical elements of their postwar constitution, conservative interests held sway in the state. Development therefore fell to individual promoters such as Jedidiah Hotchkiss and John Imboden. Hotchkiss worked with the Chesapeake and Ohio Railroad to open up coalfields in Wise County, while Imboden formed an alliance with the Norfolk and Western Railroad to develop the Flat Top Mountain coalfield. In addition to attracting capital from across the United States, these two men published promotional literature that touted the mineral wealth of the region. Imboden, as the land agent, attorney, and railroad manager for the Virginia Coal and Iron Company, drew upon his career as a former Confederate officer to coordinate the local affairs for large corporate concerns outside the Old Dominion. Outside capital with local connections characterized the development of this coal region for years to come. "Far-sighted businessmen came from the East and the North, from England and Europe, and passed through the gap into the rugged wilderness and looked at the timber and coal outcroppings and the rushing singing rivers and dreamed of industry and progress," the Virginia Coal and Iron Company later claimed, "and were so certain of their dreams coming true that they were willing to spend their time and money here.[5]

Railroads also figured prominently in the development of West Virginia's coal deposits, but Radical West Virginia's railroad policy demonstrated the inadequacy of a state-centered strategy in a national economy. As mentioned earlier, delegates to the statehood convention envisioned the B&O Railroad as a major partner in the development of the state's dormant mineral resources. But by 1869 West Virginia's relationship with railroads suggested that, despite their early optimism, the Radicals had grabbed the tiger of rapid economic development by the tail. State officials regarded the Baltimore and Ohio Railroad to be West Virginia's major commercial and industrial artery, but, when the B&O claimed it was exempt from any state taxation under their 1827 Maryland charter, this relationship soured. In March 1869 the legislature passed a joint resolution forming a committee to assess state and school taxes on the property of the Baltimore and Ohio Railroad up to 31 December 1868, which authorized the two parties to settle the matter on "fair and liberal terms."[6]

Despite government efforts, however, officials of the B&O Railroad needed little extensive lobbying in order to evade state taxes in West Virginia, as one of the state officials charged with assessing the bill on the B&O was, in fact, an employee of the railroad. Henry Gassaway Davis, a Democratic senator and a wealthy merchant who earned his fortune selling supplies to the B&O and the federal government during wartime, received an appointment on the committee to negotiate a tax settlement between West Virginia and the B&O Railroad. Although ostensibly a representative of the state, Davis won office with financial support and free railroad passes from the B&O and kept the railroad's president, John Garrett, constantly apprised of the negotiations. Davis also distributed railroad passes to state officials in order to expedite the settlement. In the end Davis's insider position played a critical role allowing the B&O to avoid paying West Virginia taxes until the 1880s.[7]

Does such a bold venture by a private corporation suggest that West Virginia's state government was "captured" by railroad interests in its earliest years? Not entirely. Because the B&O ran across several states and did not have to deal with West Virginia exclusively, the federal structure of American regulation worked against state officials. This became evident when the B&O actually retarded the growth of the northern West Virginia's coal trade through rate discrimination. By the 1850s and 1860s large railroad lines such as the B&O often used a rate structure that discriminated against short-haul in favor of long-haul traffic. Coal mining firms shipped a high-bulk and low-value product and therefore needed low rates in order to turn a profit. Because the coal region of northern West Virginia remained sparsely populated in the 1860s and coal production suffered from sporadic gluts and shortages, the B&O imposed relatively high rates on the region. Without large firms to produce a steady stream of freight and therefore command rate rebates, coal traffic originating from West Virginia remained relatively low during the 1860s. As one mine operator from the coalfields surrounding Fairmont reminisced, "it did not pay the railroad to bother with local freights," and "stopping trains to put on or off freight and to put cars in and take them out of the sidings did not pay." "Under this blighting policy," the miner speculated, "it is not strange that that the coal business in the Upper Monongahela Valley amounted to but little for many years."[8]

At the same time that the B&O undermined the authority of Radical rule in the northern counties of West Virginia, coal interests centered in Kanawha County hoped to attract a major railroad system to service their coalfields. West Virginia originally set up a Kanawha Board to oversee the improvement of water navigation along the river, but by 1869 engineering and financial materials regarding the venture remained ensconced in Richmond.[9] Southern West Virginia coal interests, tired of waiting for the improvement of the Kanawha River, pushed an incredibly liberal charter for the Chesapeake and Ohio Railroad (C&O) through the legislature in 1867. This railroad, originally chartered as the Covington and Ohio Railroad in Virginia, linked Richmond to the Blue Ridge town of Covington and sought a connection to the Ohio River by the 1860s. The C&O's West Virginia charter authorized the firm to raise thirty million dollars in capital and to hold five million acres of land in

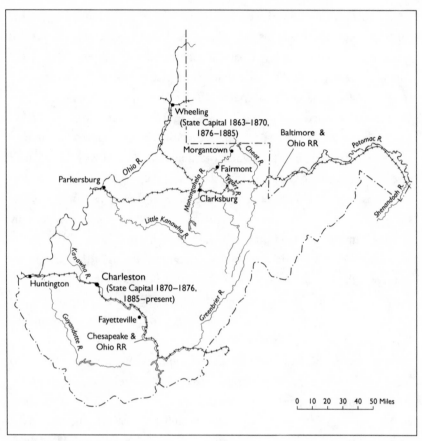

Map 5. "Radical" West Virginia

West Virginia. The C&O's charter exempted the company from taxation un-til annual net profits amounted to 10 percent of the capital invested, and Kanawha boosters skirted the constitutional restriction on stock subscrip-tions by authorizing counties to purchase stock in the C&O charter. The C&O completed its railroad line from Covington, Virginia, to Huntington, West Virginia, in 1873, finally linking the Kanawha coal region to eastern markets.[10]

The relationship of the B&O and the C&O railroads with the West Vir-ginia coal trade demonstrates the dependence of coal mining upon railroads in the postwar economy. In the northern counties thirty-five of the fifty-two coal mining firms chartered by West Virginia general laws by 1865 used the

B&O system. In the first three years of the C&O's operation twenty of the thirty-one coal mining firms created by West Virginia general laws operated along the C&O's railroad line. Coal company charters also depended upon the course of these railroads, so, after a postwar high of fifty-two charters were issued for coal companies in 1865, chartering activity in West Virginia dropped. General laws created only five coal mining companies in 1867, eight in the following year, and eleven in 1869. Many of the companies created by out-of-state investors remained absentee owners of mineral lands, and very few of them began coal mining in earnest during the 1860s. Only twenty-four of the fifty-two firms created in 1865 even bothered to open offices in West Virginia. Despite high expectations, the chief staple of West Virginia's postwar economy was a resource of limited value to the state in which it was mined.[11]

Perhaps West Virginians could draw upon the lessons of their neighbors to the North. But in Pennsylvania different challenges confronted the postbellum coal trade. On a September morning in 1869 a massive fire ripped through the Steuben Shaft in the Southern Anthracite Field at Avondale, Pennsylvania. Burning debris and smoke clogged the shaft, trapped over a hundred miners beneath the ground, and inhibited attempts to lift them to safety for twenty-four hours. Without a second outlet from the shaft, the miners at Avondale waited in vain for rescue attempts from the surface. When rescue parties finally descended the shaft the following morning, they found the bodies of 110 men bearing the clear signs of smoke inhalation and asphyxiation. As the would-be rescuers lifted the bodies of their fellow miners to the surface, thousands gathered at the shaft's mouth to mourn. Newspapers across the nation ran stories describing the carnage at Avondale, which was at that time the worst coal mining disaster in the United States.[12]

The tragedy at Avondale energized an ongoing campaign in Pennsylvania to improve safety conditions in the state's anthracite and bituminous coal fields. In 1869, only a few months before the Avondale fire, the Pennsylvania legislature had passed a law requiring basic safety standards for Schuylkill County's mines in terms of ventilation, hoisting systems, and steam engines and which created a mining inspection agent to ensure that firms complied with basic safety standards. But the legislation did not arrive in time to save the men at Avondale. In January 1870 Governor John White Geary pointed to

the Avondale disaster in his annual message to the legislature as proof that further regulatory laws were needed in order to avoid such disasters in the future.
"The lives of so useful a class of men as our miners should not, and must not,"
Geary declared, "be permitted to be thus sacrificed upon the alter of human
cupidity." The public added its voice to the governor's plea, as petitions
streamed into Harrisburg calling for a general safety act. In less than two
months the act passed through the legislature without a dissenting vote. That
same year the first report of Pennsylvania's mine inspector proved that the demand for reform was well founded when it revealed 138 violations and many
casualties due to unsafe conditions in the state's anthracite mines.[13]

The creation of mine safety laws, despite the many flaws in their execution, symbolized an altogether new relationship between public officials and
the coal trade. In Pennsylvania the business of regulating the state's coal industry in the 1870s moderated the enthusiastic promotion that characterized
the state's antebellum political economy of coal. Mine operators, especially
those in the western bituminous coal regions, had fought tooth and nail in
both the legislature and Pennsylvania's Supreme Court against safety, ventilation, and inspection provisions during the 1860s and 1870s. But by 1877 Pennsylvania had expanded the basic legislation of 1869 and 1870 to establish
statewide standards in mine safety and ventilation, along with an office of
mine inspectors. Such provisions, although effective in redressing the most
heinous of mine conditions, could not eliminate the brutal working conditions nor relieve the growing tension between operator and miner in both the
anthracite and bituminous fields. Mine inspectors, moreover, found themselves woefully understaffed and confronted by hostile and uncooperative
mine operators. They were thus unable to guarantee statewide compliance.
The actions of the Pennsylvania legislature nonetheless demonstrated the
state's willingness to get down to the business of regulating its coal trade.[14]

While ventilation and other issues of mining safety appeared in the public conscience, Pennsylvania's coal trade was undergoing big changes; it hardly
resembled the patchwork of small-scale colliers and mining corporations of
antebellum memory. Patterns of chartering and taxation begun in the turbulent Civil War years persisted in Harrisburg long after Appomattox. Pennsylvania's new constitution of 1873 forbade special chartering, but the creation of

corporations under general laws continued in earnest. So did the taxing of those corporations. In 1868 taxes on corporate stocks alone made up 21 percent of Pennsylvania's revenue. When combined with taxes on coal and railroad tonnage, the draw from corporations made up nearly a third of the state's income. At the same time, miners had little reason to cheer legislative developments beyond safety regulations. Mine operators sponsored the Coal and Iron Police Act in 1866, which created a quasi–public enforcement agency designed to break strikes and repress labor organization under the pliant guidelines of "protection of private property" and "safeguarding the interests." Even worker-friendly acts seemed to favor coal companies. In 1868 the legislature passed an eight-hour workday but included a loophole that excluded contracted workers—which included the vast majority of mine laborers—from the provisions of the act.[15]

By the 1870s new actors arrived on the stage who helped redefine the role of Pennsylvania's state government. In the anthracite region the aggressive policies of the Philadelphia and Reading's new president, Franklin Gowen, threatened to consolidate entire coalfields under a single corporate owner when he created the Philadelphia and Reading Coal and Iron Company as a holding company for his railroad. By 1875 Gowen's concerns controlled 85 percent of the Schuylkill region's anthracite collieries. When the Pennsylvania legislature summoned Gowen to justify his violation of the Philadelphia and Reading's provision against owning coal companies, he responded with an argument that highlighted the new postbellum national economy and the need for a new approach to the coal trade. "The time has passed when the legislative power of the State must say that such ruinous competition shall exist as will reduce the price of every article below the cost of its production," Gowen told the legislators. By the early 1880s his company dominated the region with very little interference from the state.[16]

Meanwhile, west of the Alleghenies, amid the rich coking fields of Pennsylvania, Henry Clay Frick proved a ruthless tactician in both accumulating smaller firms and shutting down organized labor. Coke is simply bituminous coal with impurities baked out so as to make it a more potent mineral fuel. With the discovery of the huge Connellsville field southwest of Pittsburgh, iron and steel firms began to burn coke more and more. As the region in-

creased production, Frick embarked upon an expansion program similar to
that of Franklin Gowen. By 1883 H. C. Frick and Company controlled 2,207
ovens and delivered about 45,000 tons of coke every month. When workers in
the Connellsville field went on strike in 1887, Frick and his new partner, Andrew Carnegie, broke the strike immediately. Although some Pennsylvanians
called upon their government to intervene, Frick and Carnegie continued to
consolidate their hold on the coke fields of southwestern Pennsylvania without any major opposition.[17]

Rather than focus upon the constant expansion of the Keystone State's
coal industry, Pennsylvania's legislature sought to maintain high production
levels and keep the peace in the volatile late-nineteenth-century coal trade.
Pennsylvania's Second Geological Survey, which lasted from 1874 to 1894,
brought this new attitude to state-funded geological expeditions. The initial
leader of Pennsylvania's Second Survey was J. Peter Lesley, a veteran of
Rogers's program and a noted professor of geology and mining at the University of Pennsylvania. The second survey focused upon developing detailed
maps for the state's coal, iron, and petroleum industry. Lesley, moreover, had
built his reputation and supported his family as a high-priced consultant to
these industries for the two decades prior to the second survey. Whereas
Rogers amassed reports, sketch maps, and cross-sections from his assistants,
Lesley published reports from each district as soon as they could be assembled and demanded that the reports be practical, not theoretical. Some of the
most striking maps were created for the anthracite fields, most of which were
studied between 1880 and 1889, with an eye toward reclaiming some of the
veins thought to be tapped out by the late nineteenth century. By the time it
was over the survey had spent about two-thirds of its $1.6 million dollar bud-
get on publishing—a notable contrast to Henry Darwin Rogers's more systematic and cautious pace. As a longtime consultant for private industry, Lesley knew how to deliver results that satisfied both science and capital, and the
relationship between private industry and public geologists prospered under
his watch.[18]

The politicization of mine safety issues in West Virginia would not come
until later in the nineteenth century. Rather, when members of Radical West
Virginia's legislature attempted to kick-start their coal trade with a state-

sponsored survey, they actively drew upon Pennsylvania for inspiration. Samuel Harris Daddow, an enterprising geologist from the Keystone State, wrote to Governor Arthur I. Boreman to suggest that he could lead a survey to designate West Virginia as "the first born of the free south—the richest mineral State in the Union, without exception." "By proper and united effort West Virginia," Daddow argued, "may be put forward in the march of improvement side by side with Pennsylvania." Governor Boreman bought the idea and pitched it to the legislature in his annual message of 1867, citing the "area of 15,000 square miles of coal underlying the surface of West Virginia" which required attention. The Senate transferred the authority of undertaking a survey to the state's commissioner of immigration, but the plans never got off the ground. In 1870 another geologist, John C. Stevenson of New York, argued that "the extent of these deposits is not positively ascertained, so the lands underlaid by immense deposits of mineral wealth and covered with noble timber can be bought for the merest trifle"; he warned that "many of the mining enterprises, initiated in this State, have failed through insufficient knowledge respecting the geological structure of the country." Stevenson's plan also failed to attract funding from the legislature. Ostensibly, resistance to the geological survey came from West Virginia's religious community, whose members found the implications of a geological time frame threatening to their scriptural dating of human history. In 1871 the geological survey's last gasp failed when a bill passed the House, but "several grave and reverand signeurs" in the Senate, who argued that geology was opposed to religion and cited the state's fiscal problems, killed the proposal.[19]

At the same time that Radical West Virginia's planned geological survey floundered, conservative interests in the state began calling for a rollback of the northern-influenced political structure and the reinstatement of the Old Dominion's county court system, viva voce voting, and property qualifications for suffrage. Although this conservative movement drew leadership from the state's Democratic Party, Liberal Republicans weary of Reconstruction politics also enlisted in the campaign to undermine Radical West Virginia. In 1870 Radicals lost power over the legislature, and southern conservatives immediately flexed their political muscle by moving the capital from Wheeling to Charleston. The relocation of the capital from the northern Panhandle to the heart of the Kanawha region symbolized a new direction for West Virginia—one based more upon the political traditions of the Old Dominion than that of the Unionist founders.[20]

In order to reinstate the political structure of Old Virginia, conservatives had embarked upon a campaign to restore voting privileges to former Confederates excluded under Radical West Virginia's political order. They tried to smear Radicals by identifying their reformist principles as furthering the interests of West Virginia's small African-American population in order to garner sympathy among the state's white voting population. In the words of one conservative politician in 1867, "one great object is to Identify the radicals with negro suffrage." Conservatives later adopted a more cynical, and successful, strategy. In 1869 Liberal Republicans offered the "Flick Amendment," named after its sponsor, Delegate William Flick of Pendleton County, which extended voting rights to all male citizens of West Virginia, without regard to color or prior condition. Some conservative "bitter-enders" resented the Flick Amendment's confirmation of black suffrage in West Virginia, but the move empowered the conservative movement more than it established racial equity. One conservative argued that, although the Flick Amendment enfranchised the state's African-American men, it "certainly enfranchised all white men—this is certainly a point gained, a great point in the control of elections." West Virginia's potential African-American voting population of about thirty-five hundred offered a small counterbalance to the twenty thousand former Confederate voters who returned to the state's political arena via the Flick Amendment.[21]

Once the Flick Amendment politically empowered former Confederates, Democrats and Liberal Republicans spearheaded the move for a new constitutional convention to restore the county courts as the center of governance in West Virginia. Radicals, although waning in political power, fought vehemently against a rollback of their state institutions, as they feared it would result in a further weakening of state government's authority and initiative. In 1871 Waitman Willey stumped for the Republicans against a convention. "We cannot have forgotten how in the olden time prior to the late war there was what was called a 'governing class' in Virginia," he argued. "It had its headquarters at and near Richmond. Then there was in each county an auxiliary clique chiefly composed of the old county courts." Willey charged that the courthouse clique controlled political life in the Old Dominion and "sat like a nightmare on the energies of the people, paralyzing and repressing all healthful growth and progress." Despite the best efforts of Radicals such as Willey, the reinstatement of conservative voters greased the passage of a new constitutional convention to meet in 1872.[22]

Former Confederates dominated West Virginia's constitutional convention of 1872. Samuel Price, the former lieutenant governor of Confederate Virginia, presided over the seventy-eight delegates, of whom only twelve were Republicans. Former Confederates also chaired every important committee during the convention. Conservatives pressed two issues in order to reconstruct the political structure of the Old Dominion in West Virginia. First, they unsuccessfully attempted to abolish the secret ballot and reestablish oral voting in state elections. A former Confederate veteran and delegate from Barbour County, Samuel Woods, noted that oral voting was "more manly and independent," but it also potentially placed "the poor & dependent men of the State too much under the Control, and at the mercy of the men who employ them." Oral voting had been one of the major tools in enforcing political deference to elites in antebellum Virginia, but the reluctance of conservatives to press this issue symbolized at least a begrudging acknowledgment among the delegates that a complete throwback to antebellum times would not float with their constituents.[23]

Railroad regulation constituted the second major issue, as the only sign that this convention met in the 1870s, and not the 1850s, sprung from this contentious debate. Given the conservative origins of most delegates, the amount of anti-railroad rhetoric on the floor of the convention took many observers by surprise. One delegate feared the B&O and C&O would soon control the whole state and that "there was no policy in granting liberal franchise to these big corporations to the detriment of the land-owners of the state," while another delegate argued that large corporations "were threatening to overthrow everything" and that "some such curbing was necessary." Pro-railroad delegates fought back by invoking the same pro-development rhetoric employed in the 1861 convention. One conservative delegate found an amendment reserving the right to repeal charters in certain circumstances to be "monstrous," as it "imperiled the property of corporations after they had already come among us and invested their money." Another conservative, although fully supportive of the restoration of county courts and oral voting, found anti-railroad rhetoric to be "covered all over with the hoar of antiquity." In the end railroad regulation failed to catch on in the 1872 convention, and most delegates sided with the editorialist of Charleston's *Kanawha Daily*. "Mineral lands can only be made productive and profitable through joint stock companies," the *Daily* opined. "Our hills and mountains will forever remain full of coal idle and unproductive, if we look alone to individual enterprise to develope them."[24]

The 1872 convention produced a document almost identical to Virginia's 1851 constitution. It substantially weakened the power of the governor, restored county courts as the locus of governing authority, and did virtually nothing to reform the state's impenetrable land law system. Not all West Virginians found its old-fashioned ideas to be quaint. "A Constitution should be for the whole people and not for a party," one constituent wrote to State Senator Gideon Camden in November 1872. "This new Constitution does not reflect the will of the people," the same constituent wrote to Camden later, "it's fifty years behind the age, in which we live." Out-of-state observers such as the *New York Times* described West Virginia's constitution as a "contrivance gotten up to make litigation the principle business in West Virginia—to the great impoverishment of suitors and the enrichment of the swarms of one-horse political lawyers that now feed upon the body politic." The convention's failure to reform the state's byzantine land and tax laws and its relocation of power to the county level marked the 1872 convention as West Virginia's "lawyer's convention" for many critics. Whatever their intentions, the thirty-four lawyers who made up nearly half the delegates in 1872 ensured that the future of West Virginia's coal trade owed more to the interaction of land lawyers and out-of-state corporations than to state government. By the time the constitution had passed, coal production in West Virginia barely passed one million tons. This was more than double the 484,215 tons raised in 1865 but not enough to push West Virginia into the top ranks of coal-producing states.[25]

By the 1880s the elimination of Radical West Virginia's commitment to state-fueled development was complete. The strangest element of Radical West Virginia's death is that the expansion of suffrage—long considered the panacea for the state's economic development—actually rolled back the clock on West Virginia. Radical West Virginia's plan for building a thriving industrial economy through active state intervention withered away; West Virginia's storied history with the coal industry would instead unfold at the turn of the nineteenth century. Moreover, state government would not reap the rewards of this development. Although the Constitution of 1872 included the power to tax corporations, the legislature did not exercise this authority until 1881, when it was used to tax railroad corporations. The state began taxing foreign and domestic corporations in 1885, but the system clearly favored growth over revenue. A bizarre regime grew in the absence of clear leadership from either the executive or legislative branch. In 1885 sewing machine salesmen in West Virginia paid an annual tax of twenty dollars, whereas foreign corporations paid an annual tax of fifteen dollars. By then coal production had passed three million tons per year, but the benefits of this development traveled along with the B&O and C&O lines to regions far beyond the borders of West Virginia.[26]

What does the strange story of Radical West Virginia tell us about the state-level political economy of coal? Does it teach us a lesson about corruption and regulatory capture? As previous chapters demonstrate, shady practices and dirty tricks played a large role in antebellum policy making. Corrupt practices, in fact, allowed Pennsylvania's coal industry to thrive by creating informal constraints on corporate chartering during the 1850s. These issues appear in the early history of West Virginia as well. For example, the first commissioner of immigration, Joseph H. Diss Debar, implied to the public that he could broker lucrative deals to attract immigrants if the state would repeal laws prohibiting corporations from selling land at a profit. He cited the 240,000 acres of coal land given to a Swiss agent by the Tennessee legislature, along with transportation contracts for Swiss settlers and an outright gift of 20,000 acres, and asked for "small inducements" from private landholders wishing to settle immigrants on their land. Debar blamed his agency's poor performance on the stinginess of landowners in West Virginia who would "rather live and die on the verge of personal want, than to make an apparent sacrifice" by donating land to his agency, which would then forward the titles on to needy immigrants. Debar eventually left the state for New York, where he was arrested for running a confidence scheme a few years later. The West Virginia legislature, perhaps justifiably given Debar's annoying tenure, withheld major appropriations from his successors. Corrupt venality on this scale did not kill Radical West Virginia. But it did place the state's economic future squarely in the lap of railroad officials and land lawyers.[27]

West Virginia's attempt to apply antebellum policy making to postwar conditions reveals less about the impact of corrupt practices or capture as it does about unintended consequences and the perils of growing a state's economy in an emergent national market. The nation had changed. Rather than providing a lesson in corruption and capture, the experiences of West Virginia, Virginia, and Pennsylvania's political economy of coal reveal more about confusion as the United States entered the Gilded Age. Armed only with the lessons of the past, state-level policy makers failed to control new entities such as the Baltimore and Ohio Railroad, Henry Clay Frick's empire of coke, or Franklin Gowen's sprawling network of anthracite railroads and coal companies. How could they? As Gowen himself argued to the Pennsylvania legislature, "The commercial prosperity of a commonwealth depends upon the amount of money poured into its coffers from the products which pass beyond its borders."[28]

Within this new context West Virginia's decision to revive its antebellum political structure abdicated even more authority to these entities. Although unintended, this incongruity between polity and economy created a distinctly new political economy for Gilded Age West Virginia. A belief that development—any development—of the state's resources offered the key to prosperity finally prevailed in a political atmosphere previously hostile to such activity. The campaign to encourage the rapid exploitation of mineral wealth assumed a religious fervor in which politicians eschewed regulation and taxation as the twin heresies of the gospel of development. This religion found converts north of the Mason-Dixon Line, Pennsylvania's state officials proving quiescent as Franklin Gowen and Henry Clay Frick consolidated production and power.

The nation's Coal Age had come into full fruition by 1890, when anthracite and bituminous fuel accounted for about 60 percent of the nation's energy consumption and miners raised over two tons of coal for every man, woman, and child in the United States. Mineral fuel also hastened the growth of a vibrant industrial economy that stretched across an entire continent. "Inexpensive coal underlay the developing national market," historian David Nye has argued. "Fossil fuel permitted Americans to accelerate transportation, to increase productivity, and to concentrate in the cities industry that had been dispersed in the countryside." Fire, smoke, and soot became symbols of economic prowess and civic pride for Americans, although there would be a high environmental toll to be paid later by the burning of mineral fuel. For the moment, however, "King Coal" had truly arrived. Although all three states contributed to this growing sector, the individual fortunes of Pennsylvania, Virginia, and West Virginia still sailed in different directions by the turn of the century. The old millraces had long been dry, but their impact upon old dominions and industrial commonwealths persisted well into the twentieth century.[29]

Notes

ABBREVIATIONS

APS   American Philosophical Society, Philadelphia

LCP   Library Company of Philadelphia

LOV   Library of Virginia, Richmond

PSA   Pennsylvania State Archives, Harrisburg

UVA   Alderman Library, University of Virginia, Charlottesville

VHS   Virginia Historical Society, Richmond

WVC   West Virginia Collection, West Virginia University, Morgantown

INTRODUCTION

1. Benjamin Henry Latrobe, in *The Virginia Journals of Benjamin Henry Latrobe, 1795–1798*, vol. 1: *1795–1797*, ed. Edward Carter II (New Haven: Yale University Press, 1977), 97, 145.

2. Tench Coxe, *A View of the United States* (Philadelphia: William Hall and Wrigley and Berriman, 1794), 70–71; letter of Benjamin Latrobe, in the appendix to "Report of the Secretary of the Treasury, on the Subject of Public Roads and Canals; Made in Pursuance of a Res-

olution of the Senate, of March 2, 1807," *House Reports, 17th Congress, 1st Session, 1821–1822* (Washington, D.C.: Government Printing Office, 1822), 64.

3. *United States Rail Road and Mining Register* (Philadelphia), 6 September 1856; *Register of Pennsylvania* (Philadelphia), 1 August 1829; H. Benjamin Powell, *Philadelphia's First Fuel Crisis: Jacob Cist and the Developing Market for Pennsylvania Anthracite* (University Park: Pennsylvania State University Press, 1978), 25.

4. See, for example, Fred Bateman and Thomas Weiss, *A Deplorable Scarcity: The Failure of Industrialization in the Slave Economy* (Chapel Hill: University of North Carolina Press, 1982); Frederick Siegel, *The Roots of Southern Distinctiveness: Tobacco and Society in Danville, Virginia, 1780–1865* (Chapel Hill: University of North Carolina Press, 1987); Marc Egnal, *Divergent Paths: How Culture and Institutions Have Shaped North American Growth* (New York: Oxford University Press, 1996); John Majewski, *A House Dividing: Economic Development in Pennsylvania and Virginia before the Civil War* (New York: Cambridge University Press, 2000).

5. There is a tendency at times to refer to "revolutions" in energy use. Most studies of nineteenth-century fuels suggest that changes in energy sources evolve incrementally. Although American industrialization relied mainly upon mineral fuel by the late nineteenth century, animal, wind, and waterpower also remained viable sources of energy throughout this period. For more on energy transition, see Sam H. Schurr and Bruce C. Netschert, *Energy in the American Economy, 1850–1975* (Baltimore: Johns Hopkins University Press, 1960), 31–74; Martin Melosi, "Energy Transitions in the Nineteenth-Century Economy," in *Energy and Transport: Historical Perspectives on Policy Issues,* ed. George H. Daniels and Mark H. Rose (Beverly Hills, Calif.: Sage Publications, 1982), 55–69; David E. Nye, *Consuming Power: A Social History of American Energies* (Cambridge, Mass.: MIT Press, 1998), 53–100.

6. Some economists draw from evolutionary science in order to understand the ways in which institutions create "path dependence" in economic development. See, for example, Lars Magnusson and Jan Ottosson, eds., *Evolutionary Economics and Path Dependence* (Brookfield, Vt.: Edward Elgar, 1997); Geoffrey M. Hodgson, *Evolution and Institutions: On Evolutionary Economics and the Evolution of Economics* (Brookfield, Vt.: Edward Elgar, 1999).

7. The "state-in, state-out" theory of development and its application to the American antebellum case developed during the 1960s as an example of successful industrialization in a noncommunist state. For its origins, see Carter Goodrich, "American Development Policy: The Case of Internal Improvements," *Journal of Economic History* 16 (1956): 449–60; and "State In, State Out—A Pattern of Developmental Policy," *Journal of Economic Issues* 2 (1968): 365–83. For more on the ways in which state-level political economy evolved in later decades, see sec. 1 of the essay on sources.

8. An American state of "courts and parties" is a convincing model of the federal state outlined in Stephen Skowronek, *Building a New American State: The Expansion of National Administrative Capacities, 1877–1920* (New York: Cambridge University Press, 1982). For more on the emergence of state-level authority at lower levels, see J. Mills Thornton, *Politics and Power in a Slave Society: Alabama, 1800–1860* (Baton Rouge: Louisiana State University Press, 1978);

Ballard Campbell, *Representative Democracy: Public Policy and Midwestern Legislatures in the Late Nineteenth Century* (Cambridge: Harvard University Press, 1980), L. Ray Gunn, *The Decline of Authority: Public Economic Policy and Political Development in New York, 1800–1860* (Ithaca, N.Y.: Cornell University Press, 1988); and John Larson, *Internal Improvement: National Public Works and the Promise of Popular Government in the Early United States* (Chapel Hill: University of North Carolina Press, 2001), 71–107. For more, see sec. 1 of the essay on sources.

9. Fred Bateman and Thomas Weiss offer an alternative explanation that southern industrialization was stunted but not as much as contemporaries and generations of historians have surmised. See Bateman and Weiss, *Deplorable Scarcity,* 157–63. For a more lengthy discussion of the historiography of the political economy of slavery, see sec. 1 of the essay on sources.

10. Charles Kromkowski, *Recreating the American Republic: Rules of Apportionment, Constitutional Change, and American Political Development, 1700–1870* (New York: Cambridge University Press, 2002), 36; *Miners' Journal* (Pottsville, Pa.), 31 December 1825. For more on the simultaneous expansion and limitation of suffrage during the Jacksonian Era, see Daniel Feller, *The Jacksonian Promise: America, 1815–1840* (Baltimore: Johns Hopkins University Press, 1995), 53–75; and Alexander Keyssar, *The Right to Vote: The Contested History of Democracy in the United States* (New York: Basic Books, 2000), 26–76. The literature on Jacksonian party development is huge, but, for a good overview, see Harry Watson, *Liberty and Power: The Politics of Jacksonian America* (New York: Hill and Wang, 1990).

11. Richard R. John, "Governmental Institutions as Agents of Change: Rethinking American Political Development in the Early Republic, 1787–1835," *Studies in American Political Development* 11 (Fall 1997): 348. The literature on the "Market Revolution" of the early antebellum years has grown tremendously in recent years. For a good overview, consult Charles Sellers, *The Market Revolution: Jacksonian America, 1815–1846* (New York: Oxford University Press, 1991); and Melvyn Stokes and Stephen Conway, eds., *The Market Revolution in America: Social, Political, and Religious Expressions, 1800–1880* (Charlottesville: University of Virginia Press, 1996).

12. This interpretation of antebellum Virginia politics traces its roots back to Charles Ambler, *Sectionalism in Virginia from 1776 to 1861* (1910; rpt., New York: Russell and Russell, 1964). Although the idea that eastern political leaders suppressed nascent manufacturing interests in the western counties has not seen significant revision, William Shade's latest treatment of party development in Virginia and A. Glenn Crothers's recent dissertation challenges the idea of a "backward" or "premodern" model of Virginia's political and economic landscape. See Shade, *Democratizing the Old Dominion: Virginia and the Second Party System, 1824–1861* (Charlottesville: University of Virginia Press, 1996); Crothers, "'The Projecting Spirit': Social, Economic, and Cultural Change in Post-Revolutionary Northern Virginia, 1780–1805" (Ph.D. diss., University of Florida, 1998). For a reinforcement of Ambler's basic model of sectional politics, see Richard Beeman, *The Old Dominion and the New Nation, 1788–1801* (Lexington: University Press of Kentucky, 1972); Norman K. Risjord, *Chesapeake Politics, 1781–1800* (New

York: Columbia University Press, 1978); Robert P. Sutton, *Revolution to Secession: Constitution Making in the Old Dominion* (Charlottesville: University of Virginia Press, 1989); William W. Freehling, *The Road to Disunion*, vol. 1: *Secessionists at Bay, 1776–1854* (New York: Oxford University Press, 1990), 162–96.

13. Studies concerning antebellum Pennsylvania politics often cite massive corruption and inefficiency as a by-product of the intense bargaining required to pass legislation in Harrisburg. See, for example, Louis Hartz, *Economic Policy and Democratic Thought: Pennsylvania, 1776–1860* (Cambridge: Harvard University Press, 1948); Douglas E. Bowers, "The Pennsylvania Legislature, 1815–1860: A Study of Democracy at Work" (Ph.D. diss., University of Chicago, 1974); Anne Marie Dykstra, *Region, Economy, and Party: The Roots of Policy Formation in Pennsylvania, 1820–1860* (New York: Garland, 1989).

CHAPTER 1. THE INTERSECTION OF POLITICS AND GEOLOGY

1. Thomas Ewbank, *The World a Workshop; or The Physical Relationship of Man to the Earth* (New York: D. Appleton, 1855), 73; Eli Bowen, *Coal and Coal Oil; or The Geology of the Earth, Being a Popular Description of Minerals and Mineral Combustibles* (Philadelphia: T. B. Peterson and Brothers, 1865), 477.

2. See Rondo Cameron, "A New View of European Industrialization," *Economic History Review*, 2d ser., 38 (1985): 10, 23; Carlo Bardini, "Without Coal in the Age of Steam: A Factor-Endowment Explanation of the Italian Industrial Lag before World War I," *Journal of Economic History* 57 (September 1997): 633–53.

3. J. R. Leifchild, *Our Coal and Our Coal-Pits; The People in Them, and the Scenes around Them* (London: Longman, Brown, Green and Longmans, 1856), 12; Phyllis Deane, *The First Industrial Revolution*, 2d ed. (New York: Cambridge University Press, 1979), 109–15, 136–37; A. J. Taylor, "The Coal Industry," in *The Dynamics of Victorian Business*, ed. Roy Church (Boston: Allen and Unwin, 1980), 46–62.

4. Benjamin Smith Barton to Thomas Pennant, 12 September 1792, Benjamin Smith Barton Papers, APS.

5. For a wide-ranging treatment of the significance of coal to the development of modern industry, its political nature, and its high environmental costs, see Barbara Freese, *Coal: A Human History* (Cambridge, Mass.: Perseus Publishing, 2003).

6. Robert Galloway, *A History of Coal Mining in Great Britain* (1968; rpt., New York: Augustus M. Kelley, 1882), 10; John Hatcher, *The History of the British Coal Industry*, vol. 1: *Before 1700: Towards the Age of Coal* (Oxford: Clarendon Press, 1993), 36–53, 67–69, 501–3, 548–49; Brinley Thomas, "Was There an Energy Crisis in Great Britain in the Seventeenth Century?" *Explorations in Economic History* 2 (1986): 134–52.

7. Benjamin Franklin to Dr. Igenhausz, 28 August 1785, in *Transactions of the American Philosophical Society Held at Philadelphia, for Promoting Useful Knowledge* (Philadelphia: Robert Aitken, 1786), 2:19–20.

8. Hatcher, *History of the British Coal Industry*, 187–91.

9. Michael W. Flinn, with David Stoker, *The History of the British Coal Industry,* vol. 2: *1700–1830: The Industrial Revolution* (Oxford: Clarendon Press, 1984), 114–28.

10. T. S. Ashton, *Iron and Steel in the Industrial Revolution* (Manchester, UK: Manchester University Press, 1951), 32–35; Charles K. Hyde, "The Adoption of Coke-Smelting by the British Iron Industry, 1709–1790," *Explorations in Economic History* 10 (1973): 414; and "Technological Change in the British Wrought Iron Industry, 1750–1815: A Reinterpretation," *Economic History Review,* 2d ser., 27 (1974): 190–206; Flinn, *History of the British Coal Industry,* 239–41.

11. J. H. H. Holmes, *Treatises on the Coal Mines of Durham and Northumberland: With Information Relative to the Stratifications of the Two Counties; and Containing Accounts of the Explosions from Fire-Damp, Which Have Occurred Therein for the Last Twenty Years; Their Causes; and the Means Proposed for Their Remedy, and for the General Improvements of the Mining System, by New Methods of Ventilation, &c.* (London: Baldwin, Cradock, and Joy, 1816), 70–71; Flinn, *History of the British Coal Industry,* 36–48; Peter Cromar, "The Coal Industry on Tyneside, 1771–1800: Oligopoly and Spatial Change," *Economic Geography* 53 (January 1977): 79–94; William J. Hausman, "Cheap Coals or Limitation of the Vend? The London Coal Trade, 1770–1845," *Journal of Economic History* 44 (1984): 321–28.

12. Flinn, *History of the British Coal Industry,* 329–66; J. H. Morris and L. J. Williams, "The Discharge Note in the South Wales Coal Industry, 1841–1898," *Economic History Review,* n.s. 10 (1957): 286–93.

13. *Historical Statistics of the United States, Colonial Times to 1957* (Washington, D.C.: Government Printing Office, 1960), 761; John J. McCusker and Russell R. Menard, *The Economy of British America, 1607–1789* (Chapel Hill: University of North Carolina Press, 1985), 284–85, 314–15.

14. Howard Nicholas Eavenson, *The First Century and a Quarter of American Coal Industry* (Pittsburgh: privately printed, 1942), 436; Thomas Mann Randolph to Thomas Jefferson, 3 May 1790, in *The Papers of Thomas Jefferson,* Vol. 16: *30 November 1789 to 4 July 1790,* ed. Julian Boyd (Princeton: Princeton University Press, 1961), 410.

15. "A.," "On Coal as a Fuel in America," *Literary Magazine* 2 (1804): 424; *Pennsylvania Gazette* (Philadelphia), 24 March 1790; H. E. Scudder, ed., *Recollections of Samuel Breck with Passages from His Note-Books (1771–1862)* (London: Sampson Low, Marston, Searle, and Rivington, 1877), 295.

16. *Pennsylvania Gazette,* 1 December 1784; *American Daily Advertiser* (Philadelphia), 21 December 1804; Stilwagon and Knight to Harry Heth, 12 June 1813, Harry Heth Papers, UVA; Frederick Binder, *Coal Age Empire: Pennsylvania Coal and Its Utilization to 1860* (Harrisburg: Pennsylvania Historical and Museum Commission, 1974), 8–10; H. Benjamin Powell, *Philadelphia's First Fuel Crisis: Jacob Cist and the Developing Market for Pennsylvania Anthracite* (University Park: Pennsylvania State University Press, 1978), 40–47.

17. On the ideological and practical implications of the Federalist-Republican struggle upon economic policy making, see Drew McCoy, *The Elusive Republic: Political Economy in Jeffersonian America* (New York: W. W. Norton, 1980); John R. Nelson, *Liberty and Property:*

Political Economy and Policymaking in the New Nation, 1789–1812 (Baltimore: Johns Hopkins University Press, 1987); Doron S. Ben-Atar, *The Origins of Jeffersonian Commercial Policy and Diplomacy* (New York: St. Martin's Press, 1993).

18. Carter Goodrich, "The Virginia System of Mixed Enterprise: A Study of State Planning of Internal Improvements," *Political Science Quarterly* 64 (1949): 355–87; Thomas C. Cochran, "Early Industrialization in the Delaware and Susquehanna River Areas: A Regional Analysis," *Social Science History* 1 (1977): 283–306; John Majewski, *A House Dividing: Economic Development in Pennsylvania and Virginia before the Civil War* (New York: Cambridge University Press, 2000), 13; John Larson, *Internal Improvement: National Public Works and the Promise of Popular Government in the United States* (Chapel Hill: University of North Carolina Press, 2001), 39–69.

19. See the report of Joseph Gilpen on a proposed Chesapeake and Delaware canal. See Albert Gallatin, "Report on Roads and Canals," in *American State Papers, Miscellaneous*, vol. 1: *Documents, Legislative and Executive of the Congress of the United States* (Washington, D.C.: Gales and Seaton, 1834), 760.

20. Calculation is based on Arthur Cole's estimates of the wholesale price of coal in New York City in 1812, which fluctuated between twenty-seven and twenty-eight dollars a chaldron. See Arthur Cole, *Wholesale Commodity Prices in the United States, 1700–1861* (Cambridge: Harvard University Press, 1938), 164.

21. Quote is from F. W. Taussig, *The Tariff History of the United States*, 8th ed. (New York: G. P. Putnam's Sons, 1931), 17. See also Curtis P. Nettels, *The Emergence of a National Economy, 1775–1815* (New York: Holt, Rinehart, and Winston, 1962), 324–35. Tariff level information comes from Eavenson, *First Century and a Quarter of American Coal Industry*, 12.

22. *New York Columbian*, 22 October 1811.

23. Thomas Jefferson, *Notes on the State of Virginia* (1787; rpt., Chapel Hill: University of North Carolina Press, 1955), 28; Alexander Hamilton, *Official Report on Publick Credit* (Washington, D.C.: Government Printing Office, 1790), 249–50; Tench Coxe, *A View of the United States* (Philadelphia: William Hall and Wrigley and Berriman, 1794), 180–81. Gallatin's quote is from the report of Joseph Gilpen on a proposed Chesapeake and Delaware canal. See Albert Gallatin, "Report on Roads and Canals," in *American State Papers, Miscellaneous*, vol. 1: *Documents, Legislative and Executive of the Congress of the United States* (Washington, D.C.: Gales and Seaton, 1834), 760.

24. Historians traditionally divide Virginia into four geographical sections, but the breakdown of the Old Dominion into coherent socioeconomic regions is problematic. William Shade argues convincingly that a "fourfold division fails to capture the geographical, demographic, and economic complexity of the Old Dominion"; he therefore divides the state into seven categories. I prefer to divide the state into eastern Virginia and western Virginia, given that most Virginians of the nineteenth century used this distinction, and it reflects the critical political rivalries that dominated Virginia's legislature. Shade, *Democratizing the Old Dominion: Virginia and the Second Party System, 1824–1861* (Charlottesville: University of Vir-

ginia Press, 1996), 21; Wilma Dunaway, *The First American Frontier: Transition to Capitalism in Southern America, 1700–1860* (Chapel Hill: University of North Carolina Press, 1996), 195–223.

25. Hugh Jones, *The Present State of Virginia: From Whence Is Inferred a Short View of Maryland and North Carolina* (1724; rpt., Chapel Hill: University of North Carolina Press, 1956), 144.

26. Kathleen Bruce, *Virginia Iron Manufacture in the Slave Era* (New York: Century Co., 1931), 88; *Historical Statistics of the United States*, 761.

27. Jefferson, *Notes on the State of Virginia*, 28; James T. Laing, "The Early Development of the Coal Industry in the Western Counties of Virginia, 1800–1865," *West Virginia History* 27 (January 1966): 144–45; *Niles' Register* (Baltimore), 14 November 1829.

28. John E. Stealey III, *The Antebellum Kanawha Salt Business and Western Markets* (Lexington: University Press of Kentucky, 1993), 45; Laing, "Early Development of the Coal Industry," 145.

29. Charles Ambler, *Sectionalism in Virginia from 1776 to 1861* (1910; rpt., New York: Russell and Russell, 1964), 1–99; Shade, *Democratizing the Old Dominion*, 17–49; Statistical data derived from U.S. Census Office [Second Census], *Return of the Whole Number of Persons within the Several Districts of the United States* (1802; rpt., New York: Arno Press, 1976), 69–72.

30. Robert P. Sutton, *Revolution to Secession: Constitution Making in the Old Dominion* (Charlottesville: University of Virginia Press, 1989), 50; Julian A. C. Chandler, "Representation in Virginia," *Johns Hopkins Studies in Historical and Political Science* 14 (1896): 271–73; F. Thornton Miller, *Juries and Judges versus the Law: Virginia's Provincial Legal Perspective, 1783–1828* (Charlottesville: University of Virginia Press, 1994), 12–33, 49.

31. The only restriction on this practice was that citizens could not claim two votes for the same freehold, which usually limited landholders to one vote per county. This practice remained in place until 1851. Some residents of Williamsburg and Norfolk were allowed to vote without a freehold according to more liberal provisions in their city charters. Julian A. C. Chandler, "The History of Suffrage in Virginia," *Johns Hopkins Studies in Historical and Political Science* 19 (1901): 3–18; Alexander Keyssar, *The Right to Vote: The Contested History of Democracy in the United States* (New York: Basic Books, 2000), 35–37.

32. Sutton, *Revolution to Secession*, 42–43.

33. Gordon Wood, *The Creation of the American Republic, 1776–1787* (New York: W. W. Norton, 1969), 430–67; Marc W. Kruman, *Between Authority and Liberty: State Constitution Making in Revolutionary America* (Chapel Hill: University of North Carolina Press, 1997), 35–37; Charles A. Kromkowski, *Recreating the American Republic: Rules of Apportionment, Constitutional Change, and American Political Development, 1700–1870* (New York: Cambridge University Press, 2002), 350–83.

34. *Richmond Enquirer*, 4 March 1815, 22 March 1817.

35. Duke de la Rochefoucauld Liancourt, *Travels through the United States of North America, the Country of the Iroquois, and Upper Canada, in the Years 1795, 1796, and 1797*

(London: R. Phillips, 1799), 2:62; *Richmond Enquirer,* 5 December 1815, 28 November 1817, and 4 December 1817; Samuel Paine to Richard Morris, 27 October 1801, Morris Family Papers, UVA.

36. Rochefoucauld, *Travels through the United States of North America,* 63; Ronald Lewis, *Coal, Iron, and Slaves: Industrial Slavery in Maryland and Virginia, 1715–1865* (Westport, Conn.: Greenwood Press, 1979), 65–66.

37. Lewis, *Coal, Iron, and Slaves,* 55–61; "Account of the Coal Mines in the Vicinity of Richmond, Virginia, Communicated to the Editor in a Letter from Mr. John Grammer, Jun.," *American Journal of Science* 1 (1819): 128; Harry Randolph to Harry Heth, 11 June 1812, Heth Papers, UVA.

38. Lewis, *Coal, Iron, and Slaves,* 61–64; Gerald Wilkes, *Mining History of the Richmond Coalfield of Virginia* (Charlottesville, Va.: Department of Mines, Minerals, and Energy, 1988), 9- 10, 49; B. Randolph to Harry Heth, 24 August 1810, Heth Papers, UVA.

39. Legislative Petition from Henrico County, 6 December 1803, Legislative Petition Collection, LVA.

40. John Cunliffe, Cornelius Buck, and "Henry Heath" (most likely Harry Heth) served on the board. See Act of 20 January 1802, reprinted in Samuel Shepherd, *The Statutes at Large of Virginia from October Session 1792 to December Session 1806* (Richmond: Samuel Shepherd, 1835), 2:374–77.

41. Act of 30 January 1805 reprinted in Shepherd, *Statutes at Large of Virginia,* 3:153.

42. Orris Paine to Richard Morris; 30 December 1797, 27 October 1801, and 26 November 1801, Morris Family Papers, UVA.

43. Entry of 21 May 1771, Jonathan Williams Journal, APS; Holmes, *Treatises on the Coal Mines of Durham and Northumberland,* 37–38; Flinn, *History of the British Coal Industry,* 146, 163–71, 180–89.

44. Wayland Fuller Dunaway, *History of the James River and Kanawha Company* (New York: Columbia University Press, 1922), 32n.

45. Petition from Powhatan County, 6 December 1804, Legislative Petition Collection, LVA.

46. Thomas Stevenson to Henry Heth, 12 August 1811; J. P. Pleasant to Harry Heth, 11 November 1811; John Davidson to Henry Heth, 26 September 1811; Stilwaggon and Knight to Harry Heth, 12 June 1813; Harry Heth to David Meade Randolph, 22 June 1814, Heth Papers, UVA.

47. Ida J. Lee, "The Heth Family," *Virginia Magazine of History and Biography* 42 (1934): 277; Lewis, *Coal, Iron, and Slaves,* 64–65.

48. William was in command of a merchant ship and rented a coal yard in anticipation of receiving Heth's coal that year. Harry could only send a few thousand bushels, however, and William was out of business in less than a year. William Heth to Harry Heth, 30 March 1815, 2 April 1815, 15 April 1815, Heth Papers, UVA.

49. Act of December 1800 (no day given); 20 January 1802, and 27 January 1802; reprinted

in Shepherd, *Laws of Virginia*, 2:252, 376, 386. The Allegany Turnpike later had its charter amended to include a 15 percent minimum on toll profits (387–88).

50. Act of 5 January 1804; reprinted in Shepherd, *The Statutes at Large of Virginia* (Richmond: Samuel Shepherd, 1836), 2:94–98.

51. William Pope to Harry Heth, 4 April 1812, Harry Heth Papers, UVA; Legislative Petitions from Richmond City, 13 December 1813 and 9 November 1814; Legislative Petition Collection, LVA; *Richmond Enquirer*, 2 March 1816; George Rogers Taylor, *The Transportation Revolution, 1815–1860* (New York: Holt, Rinehart, and Winston, 1951), 133.

52. J. P. Pleasants to Harry Heth, 8 July 1811; Thomas B. Main to Harry Heth, 2 December 1815, Heth Papers, UVA; Chesterfield County Petition, 15 December 1824, Legislative Petitions File, LVA.

53. Chesterfield County Petition, 23 December 1818, Legislative Petitions File, LVA.

54. The Board of Public Works, a state-run agency in Virginia designed to underwrite the cost of internal improvements, played a large role in the company's day-to-day operations. Private stockholders and directors still retained some rights, but from 1820 until 1835 the James River Company effectively operated as a state enterprise. Dunaway, *James River and Kanawha Company*, 66–68.

55. Chesterfield County Petition, 17 January 1834, Legislative Petitions File, LVA.

56. Chesterfield County Petition, 15 December 1824, Legislative Petitions File, LVA; *Laws of Virginia, 1833* (Richmond: Thomas Ritchie, 1833), 95.

57. Henrico County Petition, 18 December 1835, Legislative Petitions File, LVA.

58. The best comprehensive source for the routes, completion, and operation of early canals in this area is Christopher Baer, *Canals and Railroads of the Mid-Atlantic States, 1800–1860* (Wilmington, Del.: Regional Economic History Research Center, 1981).

59. Statistics are from "Second Auditor's Report, List of Articles Brought Down James River to Richmond," appended to the *Virginia House Journal, 1827–28* (Richmond: Thomas Ritchie, 1828); and "Summary of Statements of Receipts and Expenditures on Account of the Revenue from the Improvements under the Direction of the James River Company," appended to the *Virginia House Journal, 1829–30* (Richmond: Thomas Ritchie, 1830).

60. Dunaway, *James River and Kanawha Company*, 326.

61. Richard Brooke to Harry Heth, 7 June 1805, Harry Randolph to Heth , 1 June 1812; H. B. Christian to Heth, 1 March 1819; Heth Papers, UVA. For extended discussions on the market for hired slaves in the Virginia coal and iron industry, see Lewis, *Coal, Iron and Slaves*, 81–103; and Charles B. Dew, *Bond of Iron: Master and Slave at Buffalo Forge* (New York: W. W. Norton, 1994), 67–70.

62. Harry Heth to David Meade Randolph, 22 June 1814, Heth Papers, UVA. George Easterly of Richmond offered to construct a system to raise water but wanted a $1,000 advance and predicted that it would cost $250 to $300 worth of wages. Daniel French, a New Yorker living in Richmond, entered an agreement with Heth to pump water from the Black Heath pits for $5,000. Apparently, both of these efforts were failures by 1813. George Easterly to Harry

Heth, 3 June 1810; Agreement between Daniel French and Harry Heth, 30 November 1811; Harry Heth to Thomas Taylor, 31 January 1813, Heth Papers, UVA.

63. Oliver Evans to Harry Heth, 26 June 1813, 14 July 1813, Heth Papers, UVA; Oliver Evans to Harry Heth, 20 June 1815, Heth Family Papers, VHS.

64. Lewis, *Coal, Iron, and Slaves,* 57; Priscilla Long, *Where the Sun Never Shines: A History of America's Bloody Coal Industry* (New York: Paragon House, 1989), 38.

65. "Account of the Coal Mines in the Vicinity of Richmond, Virginia," 127; Henry Heth to Harry Heth, 24 February 1818; Harry Heth to Thomas Railey and Brother, 11 August 1816, Heth Papers, UVA; Robert Starobin, *Industrial Slavery in the Old South* (New York: Oxford University Press, 1970), 146–89; Lewis, *Coal, Iron, and Slaves,* 179–209.

66. In 1876 a mining engineer analyzed the Richmond basin and estimated that there were over one million tons of good coal remaining after nearly one hundred years of sporadic mining. See Oswald J. Heinrich, "The Midlothian, Virginia, Colliery in 1876," *Transactions of the American Institute of Mining Engineers* 4 (May 1875–February 1876): 311.

CHAPTER 2. THE COMMONWEALTH'S FUEL

1. William H. Keating, *Considerations upon the Art of Mining. To Which Are Added, Reflections on Its Actual State in Europe, and the Advantages Which Would Result from an Introduction of This Art into the United States* (Philadelphia: M. Carey and Sons, 1821), 4, 70–73.

2. Keating, *Considerations upon the Art of Mining,* 74–75, 85.

3. This is the line of argument taken in A. H. Benjamin Powell's *Philadelphia's First Fuel Crisis: Jacob Cist and the Developing Market for Pennsylvania Anthracite* (University Park: Pennsylvania State University Press, 1978), 146–48. For another argument based largely upon relative prices, see Alfred Chandler, "Anthracite Coal and the Beginnings of the Industrial Revolution in the United States," *Business History Review* 46 (Summer 1972): 153.

4. See, for example, E. W. Brian Arthur, "Competing Technologies, Increasing Returns, and Lock-In by Historical Events," *Economic Journal* 99 (March 1989): 116–31.

5. This characterization of Pennsylvania's sections is based on Philip S. Klein and Ari Hoogenboom, *A History of Pennsylvania,* 2d enl. ed. (University Park: Pennsylvania State University Press, 1980), 39–46. Anne Marie Dykstra divides Pennsylvania into at least seven distinct economic regions, but for the purposes of this brief introduction I have limited the regions to the three traditional sections of the state. See Dykstra, *Region, Economy, and Party: The Roots of Policy Formation in Pennsylvania, 1820–1860* (New York: Garland Publishing, 1989), 34–74.

6. Howard Nicholas Eavenson, *The Pittsburgh Coal Bed: Its Early History and Development* (New York: American Institute of Mining and Metallurgical Engineers, 1938), 8; Carmen DiCiccio, *Coal and Coke in Pennsylvania* (Harrisburg: Pennsylvania Historical and Museum Commission, 1996), 12–13.

7. Eavenson, *Pittsburgh Coal Bed,* 13; and *The First Century and a Quarter of the American Coal Industry* (Pittsburgh: privately printed, 1942), 165.

8. DiCiccio, *Coal and Coke in Pennsylvania,* 18–19.

9. William E. Edmunds and Edwin F. Koppe, *Coal in Pennsylvania* (Harrisburg: Pennsylvania Geological Survey, 1981), 17. Rhode Island contained a small amount of anthracite coal but not nearly enough to sustain an extensive mining industry.

10. On the early discovery of Lehigh coal, see Erskine Hazard, "History of the Introduction of Anthracite Coal into Philadelphia," *Memoirs of the Historical Society of Pennsylvania* 2 (1827): 157; John Hoffman, "Anthracite in the Lehigh Valley of Pennsylvania, 1820–1845," *United States National Museum Bulletin* 252 (1968): 92–93.

11. Israel Daniel Rupp, *History of Northampton, Lehigh, Monroe, Carbon, and Schuylkill Counties* (Harrisburg: Hickok and Cantine, 1845), 298; *Register of Pennsylvania*, 7 August 1830; James Woodhouse, "Experiments and Observations on Lehigh Coal," *Medical Museum* 1 (1805): 441–43.

12. James Swank, *History of the Manufacture of Iron in All Ages and Particularly in the United States from Colonial Times to 1891* (1892; rpt., New York: Burt Franklin, 1965), 471.

13. Powell, *Philadelphia's First Fuel Crisis*, 40–43. See also Frederick Moore Binder, *Coal Age Empire: Pennsylvania Coal and Its Utilization to 1860* (Harrisburg: Pennsylvania Historical and Museum Commission, 1974), 10; H. Benjamin Powell, "The Pennsylvania Anthracite Industry, 1769–1976," *Pennsylvania History* 47 (January 1980): 4–5.

14. Cist's original pamphlet was reprinted in the *American Journal of Science* 4 (1822): 8–16 (quote is from p. 8).

15. *American Journal of Science* 4 (1822): 10, 11, 13, 16.

16. Bruce Sinclair, *Philadelphia's Philosopher Mechanics: A History of the Franklin Institute, 1824–1865* (Baltimore: Johns Hopkins University Press, 1974), 90n; Marcus Bull, "Experiments to Determine the Comparative Qualities of Heat, Evolved in the Combustion of the Principal Varieties of Wood and Coal, Used in the United States, for Fuel; and, Also, to Determine the Comparative Quantities of Heat Lost by Ordinary Apparatus, Made Use of for Their Combustion," *Franklin Journal and American Mechanics' Magazine* 1 (May 1826): 257–93.

17. Bull, "Experiments to Determine the Comparative Qualities of Heat," 271.

18. Pennsylvania's leading coal trade publication, Pottsville's *Miners' Journal*, reprinted Bull's findings on its front page. His findings also appeared in pamphlet form and eventually circulated in European scientific circles. Binder, *Coal Age Empire*, 11.

19. Marcus Bull, *A Defence of the Experiments to Determine the Comparative Value of the Principle Varieties of Fuel Used in the United States, and Also in Europe* (Philadelphia: Judah Dobson, 1828), 35.

20. Jacob Bigelow, *A Short Reply to a Pamphlet Published at Philadelphia* (Boston: Hilliard, Gray, Little, and Wilkins, 1828), 3; Marcus Bull, *An Answer to "A Short Reply"* (Philadelphia: Judah Dobson, 1828), 8–9.

21. Benjamin Silliman, "Anthracite Coal of Pennsylvania, &c. Remarks upon Its Properties and Economical Uses," *American Journal of Science* 10 (February 1826): 331- 51; Chandos Michael Brown, *Benjamin Silliman: A Life in the Young Republic* (Princeton: Princeton University Press, 1989), 322.

22. Silliman, "Anthracite Coal of Pennsylvania," 337.

23. Silliman, "Anthracite Coal of Pennsylvania," 337, 344. Also see his letter of support in Marcus Bull, *Defence of the Experiments,* 14. Later articles focused upon both the physical characteristics of the coal and descriptions of the abundance of Pennsylvania's anthracite coal regions. See the *American Journal of Science* 13 (June 1826): 54–74, 75–83; 18 (July 1830): 308–28; 19 (January 1831): 1–21; 20 (July 1831): 163; 24 (July 1833): 173–74.

24. Bull, "Experiments to Determine," 285; *Miners' Journal,* 26 August 1826.

25. Sinclair, *Philadelphia's Philosopher Merchants,* 84.

26. *Franklin Journal and American Mechanics' Magazine* 9 (June 1830): 136; 12 (July 1831): 5–6; 13 (January 1832): 165–66; 15 (March 1833): 173–74; (April 1833): 237; 16 (July 1833): 89; (December 1833): 405–7; 421–23.

27. *Franklin Journal and American Mechanics' Magazine* 3 (February 1827): 124. See also articles in vols. 2 (November 1826): 292–96; and 10 (September 1830): 198–201.

28. Binder, *Coal Age Empire,* 19–21; *Register of Pennsylvania,* 23 February 1833.

29. Gordon Wood, *The Creation of the American Republic, 1776–1787* (New York: W. W. Norton, 1969), 83–90; Douglas M. Arnold, *A Republican Revolution: Ideology and Politics in Pennsylvania, 1776–1790* (New York: Garland Publishing, 1989), 19–24; Marc Kruman, *Between Authority and Liberty: State Constitution Making in Revolutionary America* (Chapel Hill: University of North Carolina Press, 1986), 150, 156–62; Alan W. Tully, *Forming American Politics: Ideals, Interests, and Institutions in Colonial New York and Pennsylvania* (Baltimore: Johns Hopkins University Press, 1994), 416–30.

30. The taxpaying qualification for suffrage did limit the vote to men owning property and residing in the state for one year. But, unlike Virginia, no minimum amount of property was necessary to vote, and most male Pennsylvanians qualified for legislative elections. Chilton Williamson, *American Suffrage: From Property to Democracy, 1760–1860* (Princeton: Princeton University Press, 1960), 133–34.

31. Arnold, *Republican Revolution,* 49.

32. Wood, *Creation of the American Republic,* 232–36; Arnold, *Republican Revolution,* 80–87.

33. The Radicals tried to preserve their power by creating strict loyalty oaths that disenfranchised many conservatives and by taxing wealthy citizens heavily. Dissatisfaction, however, with their attempts to use paper money to bolster a faltering economy and the successful adoption of the federal Constitution following the Philadelphia Convention of 1787 eroded the Radicals' popular support and paved the way for a new constitutional convention. See Klein and Hoogenboom, *History of Pennsylvania,* 104–7; Arnold, *Republican Revolution,* 173–232, 300–305.

34. For a more detailed discussion of the significance of Pennsylvania's constitutional revisionism in the 1780s, see Owen Ireland, "The Invention of American Democracy: The Pennsylvania Federalists and the New Republic," *Pennsylvania History* 67 (2000): 161–71.

35. Russell J. Ferguson, *Early Western Pennsylvania Politics* (Pittsburgh: University of Pitts-

burgh Press, 1938), 126–31, 252–55; Earl Thomas Bruce, "Political Tendencies in Pennsylvania," (Ph.D. diss., Temple University, 1939), 210–11; Thomas Slaughter, *The Whiskey Rebellion: Frontier Epilogue to the American Revolution* (New York: Oxford University Press, 1986), 158–71.

36. Alan W. Tully, "Ethnicity, Religion, and Politics in Early Pennsylvania," *Pennsylvania Magazine of History and Biography* 107 (1983): 491–536; Owen Ireland, "The Crux of Politics: Religion and Party in Pennsylvania," *William and Mary Quarterly* 42 (1985): 453–75; Kenneth Keller, "Cultural Conflict in Early Nineteenth-Century Pennsylvania Politics," *Pennsylvania Magazine of History and Biography* 110 (1986): 509–30.

37. Sanford W. Higginbotham, *The Keystone in the Democratic Arch: Pennsylvania Politics, 1800–1816* (Harrisburg: Pennsylvania Historical and Museum Commission, 1952), 334; James A. Kohl, *Ill Feeling in the Era of Good Feeling: Western Pennsylvania Political Battles, 1815–1825* (Pittsburgh: University of Pittsburgh Press, 1956), 28.

38. *An Act to Authorize the Governour to Incorporate a Company to Make a Lock Navigation on the River Schuylkill* (Philadelphia: Printed at the Office of the United States Gazette, 1815), 19–20; Theodore B. Klein, *The Canals of Pennsylvania and the System of Internal Improvements* (Harrisburg: Wm. Stanley Ray, 1901), 4–7, 11–12; Clifton K. Yearley, *Enterprise and Anthracite: Economics and Democracy in Schuylkill County, 1820–1875* (Baltimore: Johns Hopkins Press, 1961), 25–27; Edward Gibbons, "The Building of the Schuylkill Navigation System, 1815–1828," *Pennsylvania History* 57 (January 1990): 17.

39. *Address of the President and Managers of the Schuylkill Navigation Company, to the Stockholders* (Philadelphia: Jacob Frick and Co., 1821), 6; Gibbons, "Building of the Schuylkill Navigation System, 1815–1828," 13–43.

40. Chester Lloyd Jones, *The Economic History of the Anthracite-Tidewater Canals* (Philadelphia: University of Pennsylvania Press, 1908), 135, 151, 155.

41. Josiah White, *Josiah White's History Given by Himself* (Philadelphia: G. H. Buchanan, 1909), 74; John N. Hoffman, "Anthracite in the Lehigh Region of Pennsylvania, 1820–1845," *United States National Museum Bulletin* 252 (1968): 94, 130.

42. Engineers "improved" a river by using a series of dams and channels to provide "navigation" for vessels. A "slack water canal" system meant digging a completely new water route alongside the river. For more on the decision to convert the Lehigh system, see Hoffman, "Anthracite in the Lehigh Region," 99–101.

43. Hoffman, "Anthracite in the Lehigh Region," 119–21.

44. *Views of a Stock-Holder in Relation to the Delaware and Hudson Canal Company,* (New York: n.p., 1831), 18; "Investigator," *Exposition of the Objects and Views of the Delaware and Hudson Canal Company, in a Letter to John Bolton, Esq., President of the Company* (New York: n.p., 1831), 4–7 (both of these pamphlets can be found in the Hagley Library, Greenville, Del.). Jones, *Economic History of the Anthracite-Tidewater Canals,* 78–81.

45. Powell, "Pennsylvania Anthracite Industry, 1769–1976," 11; Yearley, *Enterprise and Anthracite,* 51–55; Hoffman, "Anthracite in the Lehigh Region of Pennsylvania," 103–6.

46. *Niles' Register,* 24 September 1825, 22 October 1825; *Register of Pennsylvania,* 20 August

1831; *Register of Pennsylvania*, 30 May 1835; Frederick Binder, "Anthracite Enters the American Home," *Pennsylvania Magazine of History and Biography* 82 (January 1958): 94.

47. Binder, *Coal Age Empire*, 16; *Niles' Register*, 8 January 1831, 29 September, 1832.

48. Statistics come from the *Register of Pennsylvania*, 2 April, 1831; and the *New York Daily Advertiser*, reprinted in the *Register of Pennsylvania*, 4 May 1833. The poem is reprinted from the *Miners' Journal*, 25 February 1832.

49. Chandler, "Anthracite Coal and the Beginnings of the Industrial Revolution," 157.

50. *Niles' Register*, 28 March 1835; "The Coal Trade of Pennsylvania," *North American Review* 42 (January 1836): 242.

51. *Pennsylvania Archives*, 4th ser., vol. 5: *Papers of the Governors, 1817–1832* (Harrisburg: W. Stanley Ray, 1900), 686.

52. *Pennsylvania Archives*, 4th ser., 5:879–86. See also Philip Shriver Klein, *Pennsylvania Politics, 1817–1832: A Game without Rules* (Philadelphia: Historical Society of Pennsylvania, 1940), 336–37.

53. *Miners' Journal*, 21 March 1830; *Pennsylvania House Journal, 1830- 31* (Harrisburg: Henry Welsh, 1831), 1:93, 130, 195, 237, 251, 269, 315, 323, 330, 341, 356, 361, 369, 381.

54. *Report of the Committee of Ways and Means, Relative to the Finances of the Commonwealth. Read in the House of Representatives, January 19, 1831* (Harrisburg: Henry Welsh, 1831), 15; *Miners' Journal*, 26 February 1831; *Pennsylvania House Journal, 1830–1831*, 1:3–5.

55. *Miners' Journal*, 17 December 1831, 6 October 1832.

56. Peter Temin, *Iron and Steel in Nineteenth Century America: An Economic Inquiry* (Cambridge, Mass.: MIT Press, 1964); W. Ross Yates, "Discovery of the Process for Making Anthracite Iron," *Pennsylvania Magazine of History and Biography* 98 (April 1974): 206–23; Darwin Stapleton, *The Transfer of Early Industrial Technologies to America* (Philadelphia: American Philosophical Society, 1987), 169–201.

57. See Yates, "Discovery of the Process for Making Anthracite Iron," 207–8.

58. *Miners' Journal*, 12 February 1831.

59. *Pennsylvania House Journal, 1837–38* (Harrisburg: Packer, Barrett, and Parke, 1838), 542–43, 854–55; *Pennsylvania Senate Journal, 1838–39* (Harrisburg: E. Guyer, 1839), 341.

60. *Laws of Pennsylvania, 1835–36* (Harrisburg: Theo. Fenn, 1836), 799–803; *Pennsylvania Archives*, 4th ser., 6:287–88; *United States Gazette*, 22 June 1836. The roll call of the vote demonstrates that the bill's main opposition came from western delegates, in whose region the charcoal iron industry remained a fixture throughout the antebellum era. From the rhetoric surrounding the bill's passage through the legislature, the coalition of delegates who voted against crippling amendments to the act, as well as the majority of firms that took advantage of the chartering opportunity, it seems that anthracite interests were the main sponsors of the legislation. See *Pennsylvania House Journal, 1835–36* (Harrisburg: Theo. Fenn, 1836), 2:1414–15; 1423–24.

61. See *United States Commercial and Statistical Register* (Philadelphia) (November 1839): 368; (April 1840): 230–31; (July 1840): 32; (November 1841): 203.

62. Walter R. Johnson, *Notes on the Use of Anthracite in the Manufacture of Iron: With Some Remarks on Its Evaporating Power* (Boston: Charles C. Little and James Brown, 1841), 32–65 (quote is from 156). According the records of Pennsylvania's secretary of state, five firms had applied for charters by 1840. This number, however, is only a rough estimate. Since the 1836 act was Pennsylvania's first general incorporation act for manufacturing corporations, record keeping was sporadic and incomplete. Entries in the ledger listing the charters listed only sixteen corporations, but contemporary and secondary sources suggest that many more incorporated iron furnaces were in operation than can be accounted for by legislative charters. For a partial list of the corporations created under the 1836 act, see Records of the Corporations Bureau, Pennsylvania State Archives, Harrisburg.

63. Quoted in the *United States Commercial and Statistical Register* (March 1841): 207.

64. Chandler, "Anthracite Coal and the Beginnings of the Industrial Revolution in the United States,"164. F. W. Taussig, the preeminent historian of the tariff, argues that high duties on iron from 1842 to 1846 helped iron makers that used charcoal fuel. Taussig, *The Tariff History of the United States* (New York: G. P. Putnam's Sons, 1903), 134–35.

65. Figures are from Temin, *Iron and Steel in Nineteenth-Century America,* app. C, 266, 269.

66. They included the Lehigh Crane Works, the Montour Iron Company, and the Shamokin Iron Company. See *Fisher's National Magazine* 3 (July 1846): 138.

67. For more on the Pennsylvania iron industry, see Paul Paskoff, *Industrial Evolution: Organization, Structure, and Growth of the Pennsylvania Iron Industry, 1750–1860* (Baltimore: Johns Hopkins University Press, 1983); and John Ingham, *Making Iron and Steel: Independent Mills in Pittsburgh, 1820–1920* (Columbus: Ohio State University Press, 1991).

68. William Pope to Harry Heth, 4 April 1812, Heth Papers, UVA; *Hunt's Merchants' Magazine* 4 (January 1841): 70.

69. *Register of Pennsylvania,* 15 November 1828; *American Journal of Science* 33 (January 1838): 200.

70. Catherine Elizabeth Reiser, *Pittsburgh's Commercial Development, 1800–1850* (Harrisburg: Pennsylvania Historical and Museum Commission, 1951), 16–20; Eavenson, *American Coal Industry,* 450.

71. Eavenson, *American Coal Industry,* 440–44.

CHAPTER 3. TRUNK AND BRANCH

1. "Proceedings of the Charlottesville Convention," appended to the *Virginia House Journal, 1828–1829* (Richmond: Thomas Ritchie, 1829), 12, 13.

2. George Rogers Taylor, *The Transportation Revolution, 1815–1860* (New York: Holt, Rinehart, and Wilson, 1951), 137; Ronald Shaw, *Erie Water West: A History of the Erie Canal* (Lexington: University Press of Kentucky, 1966), 101–22; Carol Sheriff, *The Artificial River: The Erie Canal and the Paradox of Progress, 1817–1862* (New York: Hill and Wang, 1996), 9–26.

3. This chapter focuses mostly upon publicly funded projects and their effects upon the

coal trade. For a sophisticated analysis of Virginia's failure to attract private capital juxtaposed with Pennsylvania's success, see John Majewski, "The Political Impact of Great Commercial Cities: State Investment in Antebellum Pennsylvania and Virginia," *Journal of Interdisciplinary History* 27 (Summer 1997): 1–26.

4. James T. Laing, "The Early Development of the Coal Industry in the Western Counties of Virginia, 1800–1865," *West Virginia History* 27 (January 1966): 145–48. Slavery was an important part of the Kanawha Valley's coal mining industry from its earliest origins, but the exact number of slaves and free laborers working in the trade is difficult to ascertain, given that work was both sporadic and seasonal and that slaves were often hired from owners for short periods of time. See Ronald Lewis, *Coal, Iron, and Slaves: Industrial Slavery in Maryland and Virginia, 1715- 1865* (Westport, Conn.: Greenwood Press, 1979), 46–47; John E. Stealey III, *The Antebellum Kanawha Salt Business and Western Markets* (Lexington: University Press of Kentucky, 1993), 126–29.

5. The 1816 legislation created the Virginia Board of Public Works, which supported internal improvements by subscribing to shares of turnpike, canal, and railroad companies with state funds. The initial assets of the fund were dividends received by existing state investments in internal improvement companies and banks. Future revenues from additional stock purchases as well as bonuses, dividends, and other income from Virginia's banks all were earmarked for the Board of Public Works' Internal Improvement Fund. *Revised Code of the Laws of Virginia* (Richmond: Thomas Ritchie, 1819), 201–5; Wayland Dunaway, *History of the James River and Kanawha Canal* (New York: Columbia University Press, 1922), 50–58; Carter Goodrich, "The Virginia System of Mixed Enterprise: A Study of State Planning of Internal Improvements," *Political Science Quarterly* 64 (1949): 362.

6. *Laws of Virginia, 1819–1820* (Richmond: Thomas Ritchie, 1820), 39–47; *1822–1823*, 50–58.

7. Wiley E. Hodges, "The Theoretical Basis for Anti-Governmentalism in Virginia, 1789–1836," *Journal of Politics* 9 (1947): 340–44. Robert Sutton suggests that eastern conservatives actually supported the idea of a constitutional convention by 1828 because they feared that the federal census of 1830 would support western claims of numerical superiority. See Robert Sutton, *Revolution to Secession: Constitution Making in the Old Dominion* (Charlottesville: University of Virginia Press, 1989), 68–69.

8. Julian A. C. Chandler, "The History of Suffrage in Virginia," *Johns Hopkins Studies in Historical and Political Science* 19 (1901): 21–22. Alison Goodyear Freehling points out that some eastern Virginians also agitated for suffrage reform. In Richmond, for example, about 45 percent of the adult white males could not meet the freehold requirement to vote. See Freehling, *Drift toward Dissolution: The Virginia Slavery Debate of 1831–32* (Baton Rouge: Louisiana State University Press, 1982), 39–45.

9. Sutton, *Revolution to Secession*, 80–87; Julian A. C. Chandler, "Representation in Virginia," *Johns Hopkins Studies in Historical and Political Science* 14 (1896): 33; Virginius Dabney, *Virginia: The New Dominion* (Charlottesville: University of Virginia Press, 1971), 216–18.

10. Sutton, *Revolution to Secession,* 72–102; Chandler, "Representation in Virginia," 32–44; Brenaman, *History of Virginia Conventions,* 48.

11. Stealey, *Kanawha Salt Business,* 72–73; James D. Dilts, *The Great Road: The Building of the Baltimore and Ohio, the Nation's First Railroad, 1828–1853* (Stanford, Calif.: Stanford University Press, 1993), 45.

12. Dunaway, *James River and Kanawha Company,* 86–87.

13. The Kanawha Turnpike stretched 208 miles from Barboursville to Covington and was completed at a cost of $192, 874.78. By 1830 the company had spent $91,766.72 on the improvement of the Kanawha River. Dunaway, *History of the James River and Kanawha Company,* 81–84, 88–89.

14. "Report of the Committee on Roads and Inland Navigation," House Document No. 34, 11, appended to the *Virginia House Journal, 1831–32* (Richmond: Thomas Ritchie, 1832); Stealey, *Kanawha Salt Business,* 76–88.

15. The exact amounts of state expenditures on the lower James River amounted to $640,143; on the Blue Ridge canal portion of the James, $368,401; on the Kanawha Turnpike road, $176,190.04; and on the Kanawha River improvement, $91,766.72. See "Report of the Committee of Roads and Inland Navigation," House Document No. 34, 15, appended to the *Virginia House Journal, 1831–32* (Richmond: Thomas Ritchie, 1832); *Winchester Republican,* 3 December 1830, reprinted in the *Charleston Kanawha Banner,* 17 December 1830.

16. *Virginia House Journal, 1831–32,* 131–32.

17. *Virginia House Journal, 1831–32,* 223–24; Dunaway, *History of the James River and Kanawha Company,* 99–122; "Memorial of the Citizens of Richmond," Document No. 12, appended to *Virginia House Journal, 1834–35* (Richmond: Samuel Shepherd, 1835). Kanawha interests wanted to authorize more stock purchases by the Board of Public Works or direct loans by the commonwealth in order to fund the James River and Kanawha Company's western improvements fully. See the Kanawha County Petition, 14 December 1841, Legislative Petition File, LOV.

18. These figures have been converted from bushels to tons. Charles Henry Ambler, *A History of Transportation in the Ohio Valley* (Glendale, Calif.: Arthur H. Clark, 1932), 307n.

19. "Memorial of Citizens of the County of Kanawha, Praying the Aid of the State to the James River and Kanawha Company, to Improve the Navigation of the Kanawha River" (1848), 2–3, appended to *Virginia Legislative Documents, 1848–1849* (Richmond: Samuel Shepherd, 1849).

20. "Memorial of the Citizens of the County of Kanawha" (1848), 3; Otis Rice, "Coal Mining in the Kanawha Valley to 1861: A View of Industrialization in the Old South," *Journal of Southern History* 31 (November 1965): 396–98; Laing, "Early Development of the Coal Industry," 152.

21. "Report of the Joint Committee of the Senate and House of Delegates on the Memorial of the President and Directors of the James River and Kanawha Company," appended to *Virginia House Journal, 1844–45* (Richmond: Samuel Shepherd, 1845).

22. This legislation effectively saved the James River and Kanawha Company from financial ruin by extending its deadline for completing the canal to Buchanan to 1859. See "Memorial of a Committee of the Stockholders of the James River and Kanawha Company, Asking an Extension of Their Charter of Said Company and Legislative Aid in the Farther Prosecution of the Work," appended to the *Virginia House Journal, 1846–47* (Richmond: Samuel Shepherd, 1847); *Laws of Virginia, 1846- 47* (Richmond: Samuel Shepherd, 1847), 80–84.

23. In 1839 the General Assembly considered a resolution to terminate new contracts above Lynchburg, thus effectively ending the James River and Kanawha Company line in that city. The resolution was defeated thirty-three to ninety-one on a sectional vote, suggesting that, whereas the trans-Appalachian counties and counties coterminous with the line could not pass sufficient funding for the Kanawha improvement, general support for the company existed among the majority of representatives. See *Virginia House Journal, 1838–39* (Richmond: Samuel Shepherd, 1839), 26–27; Dunaway, *History of the James River and Kanawha Company,* 153–54.

24. A sweeping study of partisan issues in antebellum state legislatures concluded that "greater variation developed within the parties than between them" in regard to internal improvements. Both Pennsylvania and Virginia were included in this study. See Herbert Ershkowitz and William G. Shade, "Consensus or Conflict? Political Behavior in the State Legislatures during the Jacksonian Era," *Journal of American History* 58 (1971): 604.

25. Henry Harrison Simms, *The Rise of the Whigs in Virginia, 1824–1840* (Richmond: William Byrd Press, 1929), 120–23; William Shade, *Democratizing the Old Dominion: Virginia and the Second Party System, 1824–1861* (Charlottesville: University of Virginia Press, 1996), 92–102,146.

26. *Thirteenth Annual Report of the President to the Stockholders of the James River and Kanawha Company, November 1847* (Richmond: James River and Kanawha Co., 1847), 21–22; *Fourteenth Annual Report of the President to the Stockholders of the James River and Kanawha Company, November 1848* (Richmond: James River and Kanawha Co., 1848), 31; "Petition of the President and Directors of the James River and Kanawha Company," appended to *Virginia Legislative Documents, 1848–49* (Richmond: Samuel Shepherd, 1849), 1, 3. The House avoided voting on the issue that provided for Kanawha improvement funding by tabling it indefinitely, and in the end the lion's share of its state funds went to improving the canal in the piedmont. In that region the James River and Kanawha Company spent roughly $3.5 million (68%) of its state authorized funds. See *Virginia House Journal, 1848–49* (Richmond: Samuel Shepherd, 1849), 376; "Report of the Committee on the Second Auditor's Report," appended to *Journal of the Acts and Proceedings of a General Convention of the State of Virginia* (Richmond: William Culley, 1850), 10–11.

27. "Memorial to the Legislature of the Commonwealth of Virginia, Adopted at Full Meeting of the Citizens of Kanawha," Document No. 8, appended to the *Virginia House Journal, 1841–42* (Richmond: Samuel Shepherd, 1842), 6–7.

28. "Report of the Select Committee on the Kanawha Memorial in Regard to the Reap-

portionment of Representation," Document No. 27, appended to the *Virginia House Journal, 1841–42*, 4; "Report of the Minority of the Select Committee to Whom Was Referred Certain Memorials Praying for a Reapportionment of Representation in Both Branches of the Legislature," Document No. 31, appended to *Virginia House Journal, 1841–42*, 3–4.

29. Local politicians George Summers and Benjamin Smith, for example, signed petitions asking the General Assembly to improve the James River and Kanawha Company. "Memorial of Citizens of the County of Kanawha, Praying the Aid of the State to the James River and Kanawha Company, to Improve the Navigation of the Kanawha River," Document No. 46, appended to *Virginia Legislative Documents, 1848–49*.

30. "Petition to the General Assembly from the Convention at Lewisburg, for the Completion of the James River and Kanawha Improvement," Document No. 7, appended to the *Virginia House Journal, 1844–1845* (Richmond: Samuel Shepherd, 1845), 2; Shade, *Democratizing the Old Dominion*, 265–66.

31. Rice, "Coal Mining in the Kanawha Valley," 407; John Barry to Christopher Quarles Tompkins, 10 August 1857, 10 April 1858, Tompkins Family Papers, Virginia Historical Society, Richmond.

32. Rice, "Coal Mining in the Kanawha Valley," 412–13; Dunaway, *History of the James River and Kanawha Company*, 197–98; J. P. Hale, "History of the Great Kanawha Slackwater Improvement," *West Virginia Historical Magazine Quarterly* 1 (April 1901): 49–75.

33. William Shade argues that, although sectionalism did play a role in Virginia politics, the state had a vibrant two-party system that demonstrated real ideological differences in the Reform Convention of 1850–51. In the end, however, sectionalism eroded the strong partisan ties that bound Whigs and Democrats together in the Old Dominion. See Shade, *Democratizing the Old Dominion*, 269–83.

34. Avard L. Bishop, "The State Works of Pennsylvania," *Transactions of the Connecticut Academy* 13 (November 1907): 165–67; George Swetnam, *Pennsylvania Transportation* (Gettysburg: Pennsylvania Historical Association, 1964), 11–42.

35. *Laws of Pennsylvania, 1823–24* (Harrisburg: Mowry and Cameron, 1824), 92–93; Bishop, "State Works of Pennsylvania," 173–74.

36. Bishop, "State Works of Pennsylvania," 182; H. Benjamin Powell, "Coal and Pennsylvania's Transportation Policy, 1825–1828," *Pennsylvania History* 38 (April 1971): 134-51.

37. The ideological split between "main liners" and "branch liners" transcended party and even region. Mathew Carey, *Brief View of the System of Internal Improvement of the State of Pennsylvania* (Philadelphia: Lydia R. Bailey, 1831), 17; Hartz, *Economic Policy and Democratic Thought*, 133–37; Ann Marie Dykstra, *Region, Economy, and Party: The Roots of Policy Formation in Pennsylvania, 1820–1860* (New York: Garland Publishing, 1989), 235–56.

38. Avard L. Bishop, "Corrupt Practices Connected with the Building and Operation of the State Works of Pennsylvania," *Yale Review* 15 (February 1907): 391–411; Hartz, *Economic Policy and Democratic Thought*, 129–80; George Rogers Taylor, *The Transportation Revolution, 1815–1860* (New York: Holt, Rinehart, and Winston, 1951), 43–45.

39. Pennsylvania's 1790 constitution provided for a septennial redistribution of the House of Representatives and the Senate according to taxable residents. Rosalind L. Branning, *Pennsylvania Constitutional Development* (Pittsburgh: University of Pittsburgh Press, 1960), 20, 26, 70–74.

40. Carey, *Brief View of the System of Internal Improvement*, 17; Douglas E. Bowers, "From Logrolling to Corruption: The Development of Lobbying in Pennsylvania, 1815–1861," *Journal of the Early Republic* 3 (Winter 1983): 439–74.

41. "Report of the Committee of Ways and Means, Relative to the Financial Concerns of the State," Document No. 249, *Pennsylvania House Journal, 1829–1830* (Harrisburg: Welsh and Miller, 1830), 2:658–659. For examples of the private coal carriers providing public inspiration, see the reports of the canal commissioners of 1832–33 and 1836–37, appended to the *Pennsylvania Senate Journal, 1832–33* (Harrisburg: Henry Welsh, 1833), 2:49–51; and the *Pennsylvania Senate Journal, 1836–37* (Harrisburg: Thompson and Clark, 1837), 2:4–5.

42. Carey, *Brief View of the System of Internal Improvement*, 25–26; Powell, "Coal and Pennsylvania's Transportation Policy, 1825–1828," 140.

43. This estimate is from Carter Goodrich, *Government Promotion of American Canals and Railroads, 1800- 1890* (1960; rpt., Westport, Conn.: Greenwood Press 1974), 65. Poor accounting and shady business practices make accurate estimates of the cost of the Main Line nearly impossible.

44. Klein, *Canals of Pennsylvania*, 15–17; Bishop, "State Works of Pennsylvania," 196- 99; Robert McCullough and Water Leuba, *The Pennsylvania Mainline Canal* (York, Pa.: American Canal and Transportation Center, 1973), 35–73.

45. For more on the anti-charter doctrine and its influence on internal improvements, see Hartz, *Economic Policy and Democratic Thought*, 69–81, 129–47.

46. "Report upon the Canal and Rail Road System," Document No. 228, *Pennsylvania House Journal, 1832–33* (Harrisburg: Henry Welsh, 1833), 2:748; Bishop, "Corrupt Practices," 411. See also Hartz, *Economic Policy and Democratic Thought*, 148–60. John Larson refers to the state works as a "desperate imitation" of the Erie Canal which used political rather than economic rationale in its design. See Larson, *Internal Improvement*, 80–87.

47. Frederick Binder, *Coal Age Empire: Pennsylvania Coal and Its Utilization to 1860* (Harrisburg: Pennsylvania Historical and Museum Commission, 1974), 116–26.

48. "*Laws of Pennsylvania, 1840* (Harrisburg: William D. Boas, 1840), 612–15; "Report of the Committee of Ways and Means, Relative to the Debts and Liabilities, and the Revenue and Resources of the Commonwealth," Document No. 84, *Journal of the House of Representatives of the Commonwealth of Pennsylvania, Session of 1843* (Harrisburg: M'Kinley and Lescure, 1843), 2:321; Bishop, "State Works of Pennsylvania," 222–23.

49. "Report of the Canal Commissioners, 1840," *Appendix to Pennsylvania House Journal, 1840* (Harrisburg: Holbrook, Henlock and Bratton, 1840), 2:7–8; "Report of the Minority of the Committee on Inland Navigation and Internal Improvement," Document No. 124, *Pennsylvania House Journal, 1840* (Harrisburg: Holbrook, Henlock, and Bratton, 1840), 2:254, 258.

50. Colleen Dunlavy, *Politics and Industrialization: Early Railroads in the United States and Prussia* (Princeton: Princeton University Press, 1994), 138–39.

51. *Pennsylvania House Journal, 1858*, 407.

52. *Niles' Register*, 13 December 1834. Howard Eavenson argues that little of the coal in the Pittsburgh area was shipped via the state works, and the data he uncovered concerning shipments *originating* at Pittsburgh substantiate this statement. The dearth of accurate data concerning the traffic, however, frustrates any attempt to quantify authoritatively coal traffic *to* Pittsburgh along the Western Division. For example, in 1839 the canal commissioners reported that 3,594 tons of coal had been shipped along the Western Division from Johnstown, but two years later they found that only 146 tons had been shipped from this point. See Eavenson, *American Coal Industry*, 186, 450, 496–97.

53. It is unclear whether this bituminous coal was actually mined in Blair County or was transported from neighboring counties. This dramatic increase in production demonstrates, however, the emergence of Hollidaysburg as a major entrepôt for the bituminous coal trade of central Pennsylvania. Eavenson, *American Coal Industry*, 486, 496; James W. Livingood, "The Canalization of the Lower Susquehanna," *Pennsylvania History* 8 (April 1941): 146–47.

54. The state charged four mills per net ton per mile, until the amount reached forty-four cents per ton, at which point the coal was allowed to pass toll-free. This toll structure lowered the price of coal carried via the Susquehanna route. See *Hunt's Merchants' Magazine* 13 (September 1845): 242–45.

55. Bishop, "State Works of Pennsylvania," 281–84.

56. Chester Lloyd Jones, *The Economic History of the Anthracite-Tidewater Canals* (Philadelphia: University of Pennsylvania Press, 1908), 63–64; John Hoffman, "Anthracite in the Lehigh Region of Pennsylvania, 1820–1845," *United States National Museum Bulletin* 252 (1968): 114.

57. Jones, *Economic History of the Anthracite-Tidewater Canals*, 65–66.

58. This report was presented by Senator Farrelly of Erie County in response to petitions from Northampton County regarding the inefficiency and high expense of the Erie Extension. See *Pennsylvania Senate Journal, 1841–42* (Harrisburg: Doas and Patterson, 1842), 1:419–28.

59. Erie's share of the coal trade in the Great Lakes from 1845 to 1856 ranged from roughly one-quarter to one-half of the net tonnage received at the major ports of Erie, Cleveland, Buffalo, Chicago, and Oswego, according to shipping statistics printed in *Hunt's Merchants' Magazine* 39 (September 1858): 383.

60. Goodrich, *Government Promotion of American Canals and Railroads*, 279.

61. Taylor, *Transportation Revolution*, 55; Goodrich, *Government Promotion of American Canals and Railroads*, 273–76; Dunlavy, *Politics and Industrialization*, 51–56; *Miners' Journal*, 3 January 1857.

CHAPTER 4. "HIDDEN TREASURES" AND NASTY POLITICS

1. J. Peter Lesley to Elizabeth Lesley, 25 April 1839; J. Peter Lesley to Sarah Allen, 26 April 1839; J. Peter Lesley to Peter Lesley, 15 May 1839, J. Peter Lesley Papers, APS.

2. *Niles' Register,* 28 November 1829.

3. Benjamin Silliman to Garrat V. Raymond, 30 June 1836 and 11 July 1836, Garret V. Raymond Correspondence, Benjamin Brand Papers, VHS; Walter B. Hendrickson, "Nineteenth Century State Geological Surveys: Early Government Support of Science," *Isis* 52 (1961): 357–71.

4. For a comprehensive account of some of the most prominent geological surveys of this time, see Anne Millbrooke, "State Geological Surveys of the Nineteenth Century" (Ph.D. diss., University of Pennsylvania, 1981).

5. *American Journal of Science* 12 (1827): 175; Millbrooke, "State Geological Surveys of the Nineteenth Century," 76–77; Patsy Gerstner, *Henry Darwin Rogers, 1808–1866: American Geologist* (Tuscaloosa: University of Alabama Press, 1994), 47–48.

6. *Niles' Register,* 2 December 1826; "Report upon the Geological Survey of the State," House Document No. 214, *Pennsylvania House Journal: 1832–33* (Harrisburg: Henry Walsh, 1833), 2:711; *American Journal of Science* 12 (1827): 176.

7. *Register of Pennsylvania,* 9 (1832).

8. The Geological Society of Pennsylvania evolved into the Association of American Geologists and Naturalists in 1840 and eight years later became the influential American Association for the Advancement of Science. Anne Millbrooke, "The Geological Society of Pennsylvania, 1832–1836," *Pennsylvania Geology* 7 (December 1976): 7–11.

9. "Report upon the Geological Survey of the State," 2:711; "Report upon the Geological and Mineralogical Survey of the State," House Document No. 194, *Pennsylvania House Journal, 1833–34* (Harrisburg: Henry Welsh, 1834), 2:848–49, 851.

10. *Pennsylvania Archives,* 4th ser., vol. 5: *Papers of the Governors, 1817–1832* (Harrisburg: William Stanley Ray, 1900), 231; *Report of the Committee Appointed on So Much of the Governor's Message as Relates to a Geological and Mineralogical Survey of the State of Pennsylvania* (Harrisburg: Theodore Penn, 1836), 4.

11. Gerstner, *Henry Darwin Rogers,* 29–30; Millbrooke, "State Geological Surveys of the Nineteenth Century," 77. On the general background of geology, see Mott T. Greene, *Geology in the Nineteenth Century: Changing Views of a Changing World* (Ithaca: Cornell University Press, 1982); Robert Muir Wood, *The Dark Side of the Earth* (London: George Allen and Unwin, 1985); and Gabriel Gohau, *A History of Geology,* trans. Albert V. Carozzi and Marguerite Carozzi (New Brunswick, N.J.: Rutgers University Press, 1990).

12. A. D. Bache and H. D. Rogers, "Analysis of Some of the Coals of Pennsylvania," *Journal of the Academy of Natural Sciences of Philadelphia* 7 (1834): 158–77; Anne Millbrooke, "Henry Darwin Rogers and the First Geological Survey of Pennsylvania," *Northeastern Geology* 3 (January 1981): 71–73.

13. Millbrooke, "State Geological Surveys," 78. The act provided for a five-year program

with an annual budget of $6,400, designed to pay for the salary of the state geologist ($2,000) and his two assistants ($1,200 each), a chemist ($1,000) and a small expense account for Rogers ($1,000). Although the lump sum averaged out to little less than Browne expected, the Pennsylvania geological survey's funding was very competitive with surveys in other states. *Laws of Pennsylvania, 1835–36* (Harrisburg: Theo. Fenn, 1836), 225–26.

14. David Hackett Fischer and James C. Kelly, *Away I'm Bound Away: Virginia and the Westward Movement* (Richmond: Virginia Historical Society, 1993), 66; Alison Goodyear Freehling, *Drift toward Dissolution: The Virginia Slavery Debate of 1831–1832* (Baton Rouge: Louisiana State University Press, 1982).

15. "Peter A. Browne to the Governor," in *Calendar of Virginia State Papers and Other Manuscripts from January 1, 1808, to December 31, 1835,* ed. H. W. Flournoy (Richmond: H. W. Flournoy, 1892), 10:588.

16. *Virginia Senate Journal, 1832–33* (Richmond: John Warrock, 1833), 10–11; *Southern Literary Messenger* 1 (November 1834): 91–93. Browne's letter was reprinted in a number of Virginia periodicals with favorable comments. For example, see Richmond's *Farmers' Register* 1 (January 1834): 504–6.

17. *Farmers' Register* 2 (August 1834): 129–34; William Barton Rogers to Henry Darwin Rogers, 30 November 1834. In Emma Rogers, ed., *Life and Letters of William Barton Rogers* (Boston: Houghton Mifflin, 1896), 1:112–13; William Ernst, "William Barton Rogers: Ante Bellum Virginia Geologist," *Virginia Cavalcade* 24 (Summer 1974): 13–21.

18. "Report from the Select Committee of the General Assembly of Virginia, to Whom Was Referred Certain Memorials from Morgan, Frederick and Shenandoah Counties, Praying for a Geological Survey of the State, with a View to the Discovery and Development of Its Geological and Mineral Resources," in William Barton Rogers, *A Reprint of Annual Reports and Other Papers, on the Geology of the Virginias* (New York: D. Appleton, 1884), 754–62; quote is from 757; Ernst, "William Barton Rogers," 15.

19. William Barton Rogers to Henry Darwin Rogers, 11 February 1835, in Rogers, *Life and Letters,* 1:116–17 (quote is from 117); Michele L. Aldrich and Alan E. Leviton, "William Barton Rogers and the Virginia Geological Survey," in *The Geological Sciences in the Antebellum South,* ed. James X. Corgan (Tuscaloosa: University of Alabama Press, 1982), 82–89; "An Act to Authorize a Geological Reconnaissance of the State, with a View to the Chemical Composition of Its Soils, Minerals, and Mineral Waters," in Rogers, *Reprint of Annual Reports and Other Papers,* 762.

20. "Report of the Geological Reconnaissance of the State of Virginia, Made under the Appointment of the Board of Public Works, 1835," in Rogers, *Reprint of Annual Reports and Other Papers,* 62, 119.

21. George Summers to William Barton Rogers, 15 February 1836, Papers of William B. Rogers, Board of Public Works Collection, LVA; Ernst, "William Barton Rogers," 15; *Laws of Virginia, 1836* (Richmond: Thomas Ritchie, 1836), 67–68; Aldrich and Leviton, "William Barton Rogers and the Virginia Geological Survey, 1835–1842," 92.

22. *Richmond Compiler and Semi-Weekly Compiler,* 16 January 1836.

23. Henry Darwin Rogers to William Barton Rogers, 23 December 1829, 27 March 1830, in Rogers, *Life and Letters,* 1:82; William Barton Rogers to Henry Darwin Rogers, 2 January 1830, *William and Mary Quarterly,* 2d ser., 7 (April 1927): 123–24 (quote is from 124).

24. The most common metaphor for mountain formation at that time was that of an ocean wave that tips over at its crest. Gerstner, *Henry Darwin Rogers,* 105–18.

25. Henry Darwin Rogers to William Barton Rogers, 10 April 1836, in Rogers, *Life and Letters,* 1:130–31; Millbrooke, "State Geological Surveys," 101.

26. Henry Darwin Rogers, *First Annual Report of the State Geologist. Read in the House of Representatives, December 22, 1836* (Harrisburg: Samuel D. Patterson, 1836), 5.

27. *Pennsylvania Archives,* 4th ser., vol. 6: *Papers of the Governors, 1832–1845* (Harrisburg: Stanley Ray, 1901), 391; Henry Darwin Rogers, *Second Annual Report on the Geological Exploration of the State of Pennsylvania. Read in the Senate, February 1, 1838* (Harrisburg: Thompson and Clark, 1838), 76, 77.

28. Gerstner, *Henry Darwin Rogers,* 87.

29. Henry Darwin Rogers, *Third Annual Report on the Geological Survey of the State of Pennsylvania. Read in the Senate, February 19, 1839* (Harrisburg: E. Guyer, 1839), 69, 96, 99; Henry Darwin Rogers, *Fourth Annual Report on the Geological Survey of the State of Pennsylvania* (Harrisburg: William D. Boas, 1840), 8, 129–70.

30. *Richmond Enquirer,* 2 February 1837; *Richmond Whig and Public Advertiser,* 3 February 1837; Henry Darwin Rogers to William Barton Rogers, 10 April 1836, in Rogers, *Life and Letters,* 1:131.

31. *Farmers' Register* 1 (July 1833): 119; Richard Sheridan, "Mineral Fertilizers in Southern Agriculture," in *The Geological Sciences in the Antebellum South,* ed. James X. Corgan (Tuscaloosa: University of Alabama Press, 1982), 73–82; and William M. Mathew, *Edmund Ruffin and the Crisis of Slavery in the Old South: The Failure of Agricultural Reform* (Athens: University of Georgia Press, 1988), 20–55.

32. The survey's first report, which was particularly keen on fertilizer, was reprinted in full in Ruffin's *Farmers' Register* 4, 1 April 1837, 713–21; "Report of the Progress of the Geological Survey of the State of Virginia for the Year 1836" and "Report of the Progress of the Geological Survey of the State of Virginia for the Year 1837," in Rogers, *Reprint of Annual Reports and Other Papers,* 126–30, 149.

33. Wallace, *St. Clair,* 203; Millbrooke, "State Geological Surveys," 98–100; Gerstner, *Henry Darwin Rogers,* 86–87.

34. J. P. Lesley to Sarah Allen, 26 April 1839, and J. P. Lesley to Peter Lesley, 25 July 1839, J. Peter Lesley Papers, APS; J. P. Lesley, *Historical Sketch of Geological Explorations in Pennsylvania and Other States* (Harrisburg: Second Geological Survey, 1876), 111.

35. *Laws of Pennsylvania, 1837–38* (Harrisburg: Henry Walsh, 1838), 380; *Register of Pennsylvania,* 28 February 1835; J. P. Lesley, *Historical Sketch of Geological Explorations,* 91.

36. "Report of Mr. Brodhead of Northampton, a Member of the Select Committee to

Whom Was Referred That Part of the Governor's Message, Relative to the Geological Survey of the State," Document No. 142, *Pennsylvania House Journal 1840–41* (Harrisburg: James S. Wallace, 1841), 444; Henry Darwin Rogers to William Barton Rogers, 25 April 1841, in Rogers, *Life and Letters,* 1:190.

37. Henry Darwin Rogers, *Sixth Annual Report on the Geological Survey of Pennsylvania* (Harrisburg: Henlock and Bratton, 1842), 6–7, 19.

38. Rogers, *Sixth Annual Report,* 24; Henry Darwin Rogers to William Barton Rogers, 22 October 1841, in Rogers, *Life and Letters,* 1:197.

39. William Barton Rogers to Henry Darwin Rogers, 22 December 1840, Board of Public Works Collection, LVA; Aldrich and Leviton, "William Barton Rogers and the Virginia Geological Survey," 94–95.

40. Charles B. Hayden to William Barton Rogers, 25 August 1838; Caleb Briggs Jr. to William Barton Rogers, 30 October 1839 and 9 September 1841, Board of Public Works Collection, LVA.

41. William Barton Rogers, "Report of the Progress of the Geological Survey of the State of Virginia for the Year 1840," in Rogers, *Reprint of Annual Reports and Other Papers,* 532–33.

42. William Barton Rogers to Henry Darwin Rogers, 8 March 1833 and 1 April 1833, in Rogers, *Life and Letters,* 1:152–53; James Brown Jr. to William Barton Rogers, 18 May 1838, Board of Public Works Collection, LVA.

43. William Barton Rogers, "Report of the Progress of the Geological Survey of the State of Virginia for the Year 1841," in Rogers, *Geology of the Virginias,* 539.

44. Anne Millbrooke, "State Geological Surveys," 110–11.

45. Gerstner, *Henry Darwin Rogers,* 129–30; 168–69; J. P. Lesley, *Historical Sketch of Geological Explorations,* 112.

46. *American Journal of Science* 41 (October 1841): 385–86; *Hunt's Merchants' Magazine* 13 (July 1845): 67; *Miners' Journal,* 22 November 1851, 18 February 1854, 7 April 1855; William B. Foulke to George S. Hart, 10 February 1851, William B. Foulke Papers, APS; Gerstner, *Henry Darwin Rogers,* 172–177.

47. Henry Darwin Rogers to William B. Foulke, 28 January 1851, William B. Foulke Papers, APS.

48. *Miners' Journal,* 8 July 1854; *Laws of Pennsylvania, 1855,* 417–18, 636–37.

49. Henry Darwin Rogers, *The Geology of Pennsylvania: A Government Survey* (Philadelphia: J. B. Lippincott, 1858), 2:13, 28.

50. Rogers, *Geology of Pennsylvania,* 540; *American Journal of Science* 28 (1859): 149–51.

51. James Brown Jr. to William Barton Rogers, 23 January 1844 and 23 February 1845, Board of Public Works Collection, LVA.

52. James Brown Jr. to William Barton Rogers, 17 February 1849, Board of Public Works Collection, LVA; Ernst, "William Barton Rogers," 20.

53. *Richmond Whig,* 11 March 1854.

54. William Barton Rogers to Henry Darwin Rogers, 11 February 1854, in Rogers, *Life and Letters*, 1:336; Ernst, "William Barton Rogers," 20.

55. C. K. Yearley Jr., *Enterprise and Anthracite: Economics and Democracy in Schuylkill County, 1820–1875* (Baltimore: Johns Hopkins Press, 1961), 103. See also Wallace, *St. Clair*, 208–15.

56. See, for example, *Report on the Lands of the Broad Top Improvement Company* (Philadelphia: James B. Chandler, 1857), 7; *United States Railroad and Mining Register*, 12 March 1870.

CHAPTER 5. MINERS WITHOUT SOULS

1. John W. Bell, *Memoirs of Governor William Smith of Virginia: His Political, Military, and Personal History* (New York: Moss Engraving Co., 1891), 9.

2. *Journal of the Senate of Virginia, 1836–37* (Richmond: John Warrock, 1837), 70–81; *Richmond Enquirer*, 5 January 1837; "An Act Prescribing General Regulations for the Incorporation of Manufacturing and Mining Companies," in *Laws of Virginia, 1836–37* (Richmond: Thomas Ritchie, 1837), 75–79.

3. *Miners' Journal*, 7 February 1838, 3 March 1838, 19 May 1838; also see "Governor Ritner's Message to the Senate Vetoing 'An Act Incorporating the Offerman Railroad and Mining Company' (April 5, 1838)," in *Pennsylvania Archives*, 4th ser., vol. 6: *Papers of the Governors, 1832–1844* (Harrisburg: Stanley Ray, State Printer, 1901), 417.

4. Anthony F. C. Wallace reconstructs the risk-taking mind-set of the small-scale mine operator in Schuylkill County and its disastrous results, in the appendix to his detailed study, *St. Clair: A Nineteenth-Century Coal Town's Experience with a Disaster-Prone Industry* (New York: Alfred A. Knopf, 1987), 446–56.

5. Daniel Raymond, *Thoughts on Political Economy* (Baltimore: Fielding Lucas, 1820), 427; *Speech of Mr. Lawrence of Belchertown, in the Senate of Massachusetts, on the Amendment Offered by Mr. Cushing to the Lowell Rail-Road Bill. February 18, 1836* (Boston: J. T. Buckingham, 1836), 13–14. For more on how historians have approached the corporation, see sec. 1 of the essay on sources.

6. Christopher Grandy, *New Jersey and the Fiscal Origins of the Modern American Corporation Law* (New York: Garland Publishing, 1993); and Charles W. McCurdy, "The Knight Sugar Decision of 1895 and the Modernization of American Corporation Law, 1869–1903," *Business History Review* 53 (Autumn 1979): 304–42.

7. George Heberton Evans, *Business Incorporations in the United States, 1800–1943* (New York: National Bureau of Economic Research, 1948), 11.

8. See *Laws of Pennsylvania, 1814–1816*. For Virginia, see the report of all taxable charters found in Document No. 14, 5–8, appended to *Virginia House Journal, 1839–40* (Richmond: Samuel Shepherd, 1840).

9. Two classic works that address the Jacksonians and the bank controversy in different ways are Arthur Schlesinger, *The Age of Jackson* (Boston: Little, Brown, 1945); and Peter Temin,

The Jacksonian Economy (New York: W. W. Norton, 1969). For a more recent treatment of antibank rhetoric, see Harry L. Watson, *Liberty and Power: The Politics of Jacksonian America* (New York: Noonday Press, 1990).

10. Charles McCool Snyder, *The Jacksonian Heritage: Pennsylvania Politics, 1833–1848* (Harrisburg: Pennsylvania Historical and Museum Commission, 1958), 76–80; James Rogers Sharp, *The Jacksonians versus the Banks: Politics in the States after the Panic of 1837* (New York: Columbia University Press, 1970), 215–46; 287–89; William Shade, *Democratizing the Old Dominion: Virginia and the Second Party System, 1824–1861* (Charlottesville: University of Virginia Press, 1996), 91–96.

11. *Pennsylvania Senate Journal, 1825* (Harrisburg: Mowry and Cameron, 1825), 304; Clifton K. Yearley Jr., *Enterprise and Anthracite: Economics and Democracy in Schuylkill County, 1820–1875* (Baltimore: Johns Hopkins Press, 1961), 89–90.

12. *Laws of Pennsylvania, 1832–1833* (Harrisburg: Henry Welsh, 1833), 167–69.

13. *Miners' Journal*, 10 March 1827. For a more detailed treatment of Schuylkill County's transformation into an anti-charter hotbed, see Yearley, *Enterprise and Anthracite*, 81–89.

14. The use of "widows and orphans" as the primary beneficiaries of corporate dividends was a common tactic employed by nineteenth-century businessmen. Josiah White, *Circular* (Harrisburg: Hamilton and Son, 1832), 1.

15. Josiah White, *To the Committee on Corporations of the Senate* (Harrisburg: Hamilton and Son, 1833), 9.

16. George Taylor, *Effect of Incorporated Coal Companies upon the Anthracite Coal Trade of Pennsylvania* (Pottsville, Pa.: Benjamin Bannan, 1833), 4, 16, 30.

17. *Pennsylvania Senate Journal, 1825*, 297, 301.

18. S. J. Packer, *Report of the Committee of the Senate of Pennsylvania upon the Subject of the Coal Trade* (Harrisburg: Henry Welsh, 1834), 19, 24.

19. Packer, *Report*, 45, 46.

20. Packer, *Report*, 56, 79.

21. On the ways in which legislators would extract money from competitors, see Douglas E. Bowers, "The Pennsylvania Legislature, 1815–1860: A Study of Democracy at Work" (Ph.D. diss., University of Chicago, 1974), 150.

22. Herbert Ershkowitz and William G. Shade, "Consensus or Conflict? Political Behavior in the State Legislatures during the Jacksonian Era," *Journal of American History* 53 (December 1971): 591–621; Bowers, "Pennsylvania Legislature," 141–42, 170–79; Ann Marie Dykstra, *Region, Economy, and Party: The Roots of Policy Formation in Pennsylvania, 1820–1860* (New York: Garland Publishing Co., 1989), 10; John N. Hoffman, "Anthracite in the Lehigh Region of Pennsylvania, 1820–1845," *United States National Museum Bulletin* 252 (Washington, D.C.: Smithsonian Institution Press, 1968), 110–11.

23. *Pennsylvania Archives*, 4th ser., vol. 6: *Papers of the Governors, 1832–1845* (Harrisburg: William Stanley Ray, 1901), 841; *Pennsylvania Senate Journal, 1845* (Harrisburg: J. M. G. Lescure, 1845), 1:693.

24. Henry W. Ruoff, ed., *Biographical and Portrait Cyclopedia of Schuylkill County, Pennsylvania, Comprising a Historical Sketch of the Country* (Philadelphia: Rush, West and Co., 1893), 213–15; Wallace, *St. Clair,* 67; Yearley, *Enterprise and Anthracite,* 88; Kevin Kenny, "Nativism, Labor, and Slavery: The Political Odyssey of Benjamin Bannan, 1850–1860," *Pennsylvania Magazine of History and Biography* 118 (1994): 325–61.

25. *Proceedings and Debates of the Convention of the Commonwealth of Pennsylvania, to Propose Amendments to the Constitution, Commenced and Held at Harrisburg, on the Second Day of May, 1837* (Harrisburg: Packer, Barrett, and Parke, 1837), 1:366, 368–69, 385.

26. Rosalind L. Branning, *Pennsylvania Constitutional Development* (Pittsburgh: University of Pittsburgh Press, 1960), 30.

27. Hartz asserts that anti-charter rhetoric in Pennsylvania, though powerful, tended to wane and wax with economic conditions and was largely displaced as a constraint upon policy during the 1850s. See Louis Hartz, *Economic Policy and Democratic Thought: Pennsylvania, 1776–1860* (Cambridge: Harvard University Press, 1948), 69–79 (quote is from 69). Statistics are from *Laws of Pennsylvania, 1834–1854;* and George Heberton Evans Jr., *Business Incorporations in the United States, 1800–1943* (New York: National Bureau of Economic Research, 1948), 12.

28. *Niles' Register,* 26 January 1828. The count of corporations comes from a report of all taxable charters by a legislative committee. See Document No. 14, 5–8, appended to *Virginia House Journal, 1839–40* (Richmond: Samuel Shepherd, 1840).

29. *Laws of Virginia, 1832–33,* (Richmond: William Ritchie, 1833), 133–36; Fletcher M. Green, "Gold Mining in Ante-Bellum Virginia," *Virginia Magazine of History and Biography* 45 (1937): 227–35; 357–66; Eugene M. Scheel, *Culpepper: A Virginia County's History through 1920* (Culpepper, Va.: Culpepper Historical Society, 1982), 126–27.

30. Legislative Petition from Henrico County, 11 December 1832, Legislative Petitions Collection, LOV.

31. Goochland County Petition, 5 January 1833, Legislative Petitions Collection, LOV.

32. Heth's firm was authorized, however, to purchase up to one thousand acres of land for timber. See *Laws of Virginia, 1832–1833,* 134. Abraham S. Wooldridge used a similar charter to form the Midlothian Coal Company from lands he already controlled in 1835. *Laws of Virginia, 1834–35,* 172–75; Chesterfield County Petition, 19 December 1834, Legislative Petitions Collection, LOV. The Cold Brook Company was authorized to sell a small portion of shares at the company's formation, but the vast majority of interest in the company remained within the Cunliffe family. See *Laws of Virginia, 1834–1835,* 175–79; *Laws of Virginia, 1839–40* (Richmond: Samuel Shepherd, 1840), 119; *1840–41* (Richmond: Samuel Shepherd, 1841), 151.

33. "Report of the Committee Appointed to Examine Charters," Document No. 14, appended *to Virginia House Journal, 1839–1840* (Richmond: Samuel Shepherd, 1839), 5–6; *Richmond Enquirer,* 7 January 1837; *Richmond Whig,* 10 February 1837.

34. The quotation from *Dartmouth College v. Woodward* is from Stanley Kutler, ed., *The Supreme Court and the Constitution: Readings in American Constitutional History* (New York:

W. W. Norton, 1984), 74; *Rider v. the Nelson and Albemarle Union Factory,* in Benjamin Watkins Leigh, *Reports of Cases Argued and Determined in the Court of Appeals, and in the General Court of Virginia* (Richmond: Shepherd and Colin, 1838), 7:154–57 (quote from Judge Tucker is on 156–57).

35. "An Act Prescribing General Regulations for the Incorporation of Manufacturing and Mining Companies," in *Laws of Virginia, 1836–37* (Richmond: Thomas Ritchie, 1837), 75–79. The breakdown of voting on these amendments suggests that Smith represented an anticorporate faction of the Democratic Party, while Whigs voted in favor of the act. Yet just as many, or more, Democrats opposed Smith, which demonstrates the intraparty split of the Virginia Democracy on this issue. See *Journal of the Senate of Virginia, 1836–37* (Richmond: John Warrock, 1837), 70–81.

36. Howard Nicholas Eavenson, *The First Century and a Quarter of American Coal Industry* (Pittsburgh: privately printed, 1942), 400; Thomas Senior Barry, *Western Prices before 1861: A Study of the Cincinnati Market* (Cambridge: Harvard University Press, 1943), 276–79.

37. By 1865 the number of steam engines used by Pennsylvania's anthracite industry reached nearly eight hundred and exceeded the power of all the stationary steam power in the United States reported in 1838—a further testament to the capital investment required for mining anthracite by the mid-nineteenth century. Louis Hunter and Lynwood Bryant, *A History of Industrial Power in the United States, 1780–1930,* vol. 3: *The Transmission of Power* (Cambridge, Mass.: MIT Press, 1991), 418–19.

38. Eli Bowen, *The Coal Regions of Pennsylvania, Being a General, Geological, Historical, and Statistical Review of the Anthracite Coal Districts Illustrated with Colored Maps and Engravings, and Containing Numerous Statistical Tables* (Pottsville, Pa.: E. N. Carvalho and Co., 1848), 34; Miners' Journal, 1 April 1848, 24 February 1849. Some historians share the view of contemporaries that the shift from special to general charters reflected a growing acceptance of corporations by Americans and ushered in an era of more egalitarian chartering. See L. Ray Gunn, *The Decline of Authority: Public Economic Policy and Political Development in New York, 1800–1860* (Ithaca, N.Y.: Cornell University, 1988), 222–45. For this interpretation of Pennsylvania's use of general incorporation, see Dykstra, *Region, Economy, and Party,* 224–25.

39. *Laws of Pennsylvania, 1849* (Harrisburg: Theo. Fenn, 1849), 563–69.

40. The various acts and supplements are found in *Laws of Pennsylvania, 1849,* 563; *1851* (Harrisburg: Theo. Fenn, 1851), 577; *1852* (Harrisburg: A. Boyd Hamilton, 1852), 496, 624; *1853* (Harrisburg: A. Boyd Hamilton, 1853), 269–70, 518, 637–38, 673; *1854* (Harrisburg: A. Boyd Hamilton, 1854), 215; *1856* (Harrisburg: A. Boyd Hamilton, 1856), 7; *1860* (Harrisburg: A. Boyd Hamilton, 1860), 39–40, 343, 629. Statistics on chartering are from the Records of the Department of State, Corporations Bureau, Letters Patent, 1814–74, PSA; Governor William Bigler from his annual message to the Assembly (4 January 1854), from *Pennsylvania Archives,* 4th ser., vol. 7: *Papers of the Governors, 1845–1858* (Harrisburg: William Stanley Ray, 1902), 649.

41. *Laws of Pennsylvania, 1854,* 437–39; *1856,* 283; *1857* (Harrisburg: A. Boyd Hamilton, 1857), 199; *1862,* 403.

42. *Pennsylvania Archives,* 4th ser., vol. 7: *Papers of the Governors, 1845–1858* (Harrisburg: William Stanley Ray, 1902), 884; *Miners' Journal,* 24 March 1855.

43. See, for example, *Report on the Coal Lands of the Zerbe's Run and Shamokin Improvement Company . . . with the Charter* (Boston: Thurston, Torry and Co., 1850), 22–24; *Charter and By-Laws of the Susquehanna Coal Company* (Philadelphia: n.p., 1851); *Charter, Coal Fields, Family Plan . . . of the Summit Branch Railroad Company* (Philadelphia: King and Baird, 1857), 22–32; *Charter and By-Laws of the Short Mountain Coal Company* (Baltimore: John F. Wiley, 1859).

44. Records of the Department of State, Corporations Bureau, Letters Patent, 1814–74, PSA.

45. *Laws of Pennsylvania, 1832–1833,* 167–70; Records of the Department of State, Corporations Bureau, Letters Patent, 1814–74, PSA, 19 December 1855, 12 February 1856, and 16 March 1857; *United States Railroad and Mining Register,* 5 July 1856.

46. Records of the Department of State, Corporations Bureau, Letters Patent, 1814–74, PSA.

47. *Miners' Journal,* 6 March 1852.

48. An example of the broad powers often bestowed upon improvement companies can be seen in the example of the Swatara Company, which became a boilerplate charter for improvement companies and is representative of the vague boundaries implicit in these special acts. See *Laws of Pennsylvania, 1849,* 126–27.

49. Records of the Corporations Bureau, Letters Patent, 1814–74, PSA; Dykstra, *Region, Economy, and Party,* 225.

50. *Miners' Journal,* 17 November 1827.

51. Ronald Lewis, *Coal, Iron, and Slaves: Industrial Slavery in Maryland and Virginia, 1715–1865* (Greenwood, Conn.: Greenwood Press, 1979), 197–209.

52. Kanawha County Petition, 6 February 1835 (1) and (2), Legislative Petitions Collection, LOV; John E. Stealey, *The Antebellum Kanawha Salt Business and Western Markets* (Lexington: University Press of Kentucky, 1993), 85–88; 158–69; 191–97.

53. *Laws of Virginia, 1839–40* (Richmond: Samuel Shepherd, 1841), 122; *1847–48* (Richmond: Samuel Shepherd, 1848), 313; *Miners' Journal,* 17 May 1851; *Cannel-Coal Company, of Coal River, Virginia* (New York: Baker, Godwin and Co., 1851); Henry A. Du Bois, *A Statement in Regard to the Coal Fields of the Virginia Cannel Coal Company, Situated in Boone County, State of Virginia, United States of America* (London: Woodfall and Kinder, 1854); Otis K. Rice, "Coal Mining in the Kanawha Valley to 1861: A View of Industrialization in the Old South," *Journal of Southern History* 31 (1965): 396–98.

54. Fletcher M. Green, *Constitutional Development in the South Atlantic States, 1776–1860* (1930; rpt., New York: W. W. Norton, 1966), 293–94.

55. *Richmond Daily Dispatch,* 1 March 1854.

56. Catlett's remarks summarized in the *Virginian* (Lynchburg), 6 March 1854; John Prosser Tabb to Charles Quarles Tompkins, 20 November 1855, Tompkins Family Papers, VHS.

57. Robert P. Sutton describes Virginia's governor as having "neither power nor dignity" throughout the entire antebellum period and asserts that, up until 1861, Virginia's legislative and executive branch of government had no real separation of powers. See Sutton, *Revolution to Secession*, 152.

58. *Virginia House Journal, 1852–53* (Richmond: John Warrock, 1853), 13, 244, 312; *Virginia Senate Journal, 1853* (Richmond: William F. Ritchie, 1853), 83, 166, 210; *Laws of Virginia, 1853–54* (Richmond: William Ritchie, 1854), 32–33.

59. *Eastern* in this case denotes east of the Blue Ridge. *Western* includes the Shenandoah Valley and what eventually became southwestern Virginia and West Virginia. See *Virginia Senate Journal, 1853–54* (Richmond: John Warrock, 1853), 333.

60. A. G. Roeber, *Faithful Magistrates and Republican Lawyers: Creators of Virginia Legal Culture, 1680–1810* (Chapel Hill: University of North Carolina Press, 1981), 260; F. Thornton Miller, *Juries and Judges versus the Law: Virginia's Provincial Legal Perspective, 1783–1828* (Charlottesville: University of Virginia Press, 1994), 113–20.

61. *Speeches of Waitman T. Willey of Monongalia County before the State Convention of Virginia, on the Basis of Representation; on County Courts and County Organization, and on the Election of Judges by the People* (Richmond: William Culley, [1851]), 27.

62. The occupational breakdown of the delegates occurs in the *Lynchburg Virginian*, 23 February 1854. Pennsylvania's lower house during the 1850s was composed of 36 percent farmers and 22 percent lawyers; it is interesting to note that Virginia's numbers appear much more similar to those of Pennsylvania during the 1820s, when farmers made up 52 percent of the lower house and lawyers only 13 percent. See Bowers, "Pennsylvania Legislature, 1815–1860," 25–26.

63. "Mining: Its Embarrassments and Its Results," *Mining Magazine* 2 (June 1854): 637–38.

64. This estimate of one million dollars in capitalization is a high one, as most coal mining firms in Pennsylvania were limited to less than a million dollars in initial capitalization. Nevertheless, the significantly low capitalization of the circuit court charters represented a huge barrier. See "Mining in Wall Street," *Mining Magazine* 4 (April 1855): 372; *Laws of Virginia, 1855–56*, 33–34; "Communication from the Secretary of the Commonwealth Enclosing [a] List of Companies Incorporated by the Courts of this State," Senate Document No. 2, 6, appended to *Virginia Senate Documents, 1857–58* (Richmond: John Warrock, 1858).

65. "Is Mining a Legitimate Business?—Mines as a Means of Investment," *Mining and Statistic Magazine* 10 (May 1858): 375.

66. Geographer Michael J. Webber argues that these "economies of agglomeration" play a crucial role in determining industry location and growth. See Webber, *Industrial Location* (Beverly Hills, Calif.: Sage Publications, 1984), 78–79. Michael Storper and Richard Walker also make a strong case for endogenous factors in locational theory when they assert that "industries are capable of generating their own conditions of growth in place, by making factors of production come to them or causing factor supplies to come into being where they did not

exist before." See Storper and Walter, *The Capitalist Imperative: Territory, Technology, and Industrial Growth* (Oxford: Basil Blackwell, 1989), 71.

67. See appendix to *Virginia Senate Journal, 1857–58*.

68. *United States Mining and Railroad Register*, 10 July 1859; D. T. Ansted, *Scenery, Science, and Art; Being Extracts from the Note-Book of a Geologist and Mining Engineer* (London: John Van Voorst, 1854), 272; John Barry to Christopher Quarles Tompkins, 4 January 1859, Tompkins Family Papers, VHS.

CHAPTER 6. THREE SEPARATE PATHS

1. Granville Davisson Hall, *The Rending of Virginia, a History* (Chicago: Mayer and Miller, 1902), 34, 38.

2. For an interpretation of the American Civil War as the "last capitalist revolution" in a comparative perspective, see Barrington Moore, *Social Origins of Dictatorship and Democracy: Lord and Peasant in the Making of the Modern Worlds* (Boston: Beacon Press, 1966), 111–55; and Raimondo Luraghi, "The Civil War and the Modernization of American Society," *Civil War History* 18 (1972): 230–50. For a dissenting view, see William Shade, "'Revolutions May Go Backward': The American Civil War and the Problem of Political Development," *Social Science Quarterly* 55 (December 1974): 753–67. Recent works such as James Huston, "Property Rights in Slavery and the Coming of the Civil War," *Journal of Southern History* 65 (May 1999): 249–86; and Marc Egnal, "The Beards Were Right: Parties in the North, 1840–1860," *Civil War History* 47 (2001): 30–56, place economic questions at the center of this debate.

3. *Miners' Journal*, 29 July 1865.

4. John F. Coleman, *The Disruption of the Pennsylvania Democracy, 1848–1860* (Harrisburg: Pennsylvania Historical and Museum Commission, 1975), 61–79; Erwin Stanley Bradley, *The Triumph of Militant Republicanism: A Study of Pennsylvania and Presidential Politics, 1860–1872* (Philadelphia: University of Pennsylvania Press, 1964), 54–96.

5. Avard L. Bishop, "The State Works of Pennsylvania," *Transactions of the Connecticut Academy* 13 (November 1907): 256; James A. Ward, *J. Edgar Thomson: Master of the Pennsylvania* (Westport, Conn.: Greenwood Press, 1980), 111. For more on the controversy over the tonnage tax's repeal and its implications for laissez-faire ideology, see Louis Hartz, *Economic Policy and Democratic Thought: Pennsylvania, 1776–1860* (Cambridge: Harvard University Press, 1948), 267–85.

6. *Laws of Pennsylvania, 1861* (Harrisburg: A. Boyd Hamilton, 1861), 88–94; Samuel R. Kamm, "The Civil War Career of Thomas Scott," (Ph.D. diss., University of Pennsylvania, 1940), 13–18; Ward, *J. Edgar Thomson*, 116; Bradley, *Triumph of Militant Republicanism*, 143–44.

7. Ward, *J. Edgar Thomson*, 116; Stanton Ling Davis, *Pennsylvania Politics, 1860–1863* (Cleveland: Western Reserve University Library Press, 1935), 199–212; Bradley, *Triumph of Militant Republicanism*, 146–49.

8. Bradley, *Triumph of Militant Republicanism*, 128–37; Keith Edward Wagner, "Economic

Development in Pennsylvania during the Civil War, 1861–1865" (PhD. diss., Ohio State University, 1969), 216–17.

9. Although it clearly boosted coal production, the general impact of the war upon the industrial economy of the northern states such as Pennsylvania is difficult to gauge accurately. Most historians concur that wartime demand encouraged very few lasting innovations or long-term increases in productivity in the North's economy during the war. See Victor Clark, "Manufacturing Development during the Civil War," *Military Historian and Economist* 3 (1918): 92–100; Emerson David Fite, *Social and Industrial Conditions in the North during the Civil War* (New York: Macmillan, 1910), 96–97; Claudia Goldin and Frank Lewis, "The Economic Cost of the American Civil War: Estimates and Implications," *Journal of Economic History* 35 (1975): 299–326; Saul Engelbourg, "The Economic Impact of the Civil War on Manufacturing Enterprise," *Business History* 21 (1979): 157; Thomas C. Cochran, "Did the Civil War Retard Industrialization?" *Mississippi Valley Historical Review* 48 (1961): 197–210; and Stephen Salsbury, "The Effect of the Civil War on American Industrial Development," in *The Economic Impact of the American Civil War*, ed. Ralph Andreano (Cambridge, Mass.: Schenkman Publishing, 1962); Patrick K. O'Brien, *The Economic Effects of the American Civil War* (Atlantic Highlands, N.J.: Humanities Press International, 1988); Roger Ransom, "The Economic Consequences of the American Civil War," in *The Political Economy of War and Peace*, ed. Murray Wolfson (Boston: Kluwer Academic Publishers, 1998), 49–74.

10. In his study of wartime Philadelphia, J. Matthew Gallman found an economic "pattern of confusion and experimentation followed by adjustment, organization, and improved efficiency." Gallman, *Mastering Wartime: A Social History of Philadelphia during the Civil War* (New York: Cambridge University Press, 1990), 292.

11. The count of legislation comes from the session laws printed annually under the title *Laws of Pennsylvania*. Occasionally, a law passed in one session will not show up until subsequent sessions, so the hand count of laws passed in the session is an approximation. The reporter's quote is from the *Miners' Journal*, 2 April 1864.

12. Charles Lyman to George Magee, 20 January 1863, 21 January 1862, 6 February 1863, 8 February 1863, Records of the Fall Brook Coal Company, PSA.

13. John Ewen to Eli Slifer, 5 March 1861, Slifer-Dill Papers, PSA.

14. *United States Railroad and Mining Register*, 28 August 1864; *Pennsylvania House Journal, 1864* (Harrisburg: Singerly and Myers, 1864), 36–37.

15. *Laws of Pennsylvania, 1863* (Harrisburg: Singerly and Myers, 1863), 1102–9; *Pennsylvania Senate Journal, 1863* (Harrisburg: Singerly and Myers, 1863), 567.

16. *Pennsylvania Senate Journal, 1863*, 567. The count of coal mining firms created by the 1863 general chartering act comes from the Pennsylvania Corporations Bureau, Letters Patent, 1863–74, Records of the Department of State, PSA.

17. Curtain mentioned his regret during his annual message to the Pennsylvania legislature in January 1865. *Pennsylvania Archives*, 4th ser., vol. 8: *Papers of the Governors, 1858–1871* (Harrisburg: W. Stanley Ray, 1902), 653.

18. *Miners' Journal*, 29 July 1865; *United States Railroad and Mining Register*, 25 April 1863.

19. *United States Railroad and Mining Register*, 27 February 1864; *Miners' Journal*, 26 March 1864, 16 April 1864.

20. Samuel Harries Daddow, *Coal, Iron and Oil; or, the Practical American Miner* (Philadelphia: J. B. Lippincott, 1866), 705, 765, 767–75; Records of the State Corporations Bureau, Letters Patent, 1814–74, PSA; Yearley, *Enterprise and Anthracite*, 7; Wallace, *St. Clair*, 72–75, 118–19, 406–7.

21. *Miners' Journal*, 23 January 1864, 21 January 1865, 20 January 1866.

22. The six firms were the Powelton Coal and Iron Company (1861), the Kittanning Coal Company (1863), the Moshannon Coal Company (1864), the Union Coal Company (1865), the Derby Coal Company (1866), and the Enterprise Coal Company (1867). James Macfarlane, *The Coal-Regions of America: Their Topography, Geology, and Development* (New York: D. Appleton, 1875), 662; Howard Nicholas Eavenson, *The First Century and a Quarter of American Coal Industry* (Pittsburgh: privately printed, 1942), 453.

23. Macfarlane, *Coal-Regions of America*, 662.

24. George H. Thurston, *Directory of the Monongahela and Youghiogheny Valleys: Containing Brief Historical Sketches of the Various Towns Located on Them; with a Statistical Exhibit of the Collieries upon the Two Rivers* (Pittsburgh: A. A. Anderson, 1859), 253–66; Macfarlane, *Coal-Regions of America*, 664–65.

25. The two firms asking for railroad privileges were the Howard Coal and Iron Company and the Madera Coal and Improvement Company. *Pennsylvania Archives, Papers of the Governors, 1858–1871*, 736–38. *Miners' Journal*, 19 March 1864; Bogen, *Anthracite Railroads*, 47–49; *Laws of Pennsylvania, 1861* (Harrisburg: A. Boyd Hamilton, 1861), 410–11. The 1869 legislation prohibited railroad and canal companies from purchasing stocks or bonds of mining firms in Schuylkill County. *Laws of Pennsylvania, 1869* (Harrisburg: State Printer, 1869), 31–32; Wagner, "Economic Development in Pennsylvania during the Civil War," 119.

26. *Pennsylvania Archives, Papers of the Governors, 1858–1871*, 523; *Important General Laws, Passed at the Session of 1864; Together with the Supplements Thereto, Passed at the Extra Session* (Harrisburg: Singerly and Myers, 1864), 3–6; "Auditor General's Report for 1865," *Pennsylvania Executive Documents, 1865* (Harrisburg: Singerly and Myers, 1865), 5.

27. Additional sources of revenue for the state included licensing fees, inheritance taxes, and other small taxes on business and legal transactions. The data for state revenues in each year are drawn from the auditor general's report for 1860–66, which can be found appended to the bound set of *Pennsylvania Executive Documents* for each respective year.

28. Grace Palladino, *Another Civil War: Labor, Capital, and the State in the Anthracite Regions of Pennsylvania, 1840–68* (Urbana: University of Illinois Press, 1990), 43–48.

29. J. Walter Coleman, *Labor Disturbances in Pennsylvania, 1850–1880* (Washington, D.C.: Catholic University of America Press, 1936), 23–25; Harold W. Aurand, *From the Molly Maguires to the United Mine Workers: The Social Ecology of an Industrial Union, 1869–1897* (Philadelphia: Temple University Press, 1971), 66–67.

30. Yearley, *Enterprise and Anthracite*, 165–67; Palladino, *Another Civil War*, 60–63.

31. For more on the work routines of nineteenth-century anthracite miners, see Priscilla Long, *Where the Sun Never Shines: A History of America's Bloody Coal Industry* (New York: Paragon House, 1989), 24–51.

32. Palladino, *Another Civil War*, 126–28 (quote is from 128).

33. Coleman, *Labor Disturbances in Pennsylvania, 1850–1880*, 40–49; Arnold Shankman, "Draft Riots in Civil War Pennsylvania," *Pennsylvania Magazine of History and Biography* 101 (1977): 190–204; Palladino, *Another Civil War*, 140–62; Kevin Kenny, *Making Sense of the Molly Maguires* (New York: Oxford University Press, 1998), 87–102.

34. Alexander Trachtenberg, *The History of Legislation for the Protection of Coal Miners in Pennsylvania, 1824–1915* (New York: International Publishers, 1942), 12–16; Wallace, *St. Clair*, 276–81; *Pennsylvania Archives, Papers of the Governors, 1858–1871*, 466–67, 544–45.

35. Charles B. Dew, *Ironmaker to the Confederacy: Joseph R. Anderson and the Tredegar Iron Works* (New Haven: Yale University Press, 1966), 34.

36. Bruce, *Virginia Iron Manufacture*, 406–9; Dew, *Ironmaker to the Confederacy*, 149–50. Mine labor, however, was almost always in short supply. Tredegar struggled mightily to keep enough free and slave miners at work during the war. Dew, *Ironmaker to the Confederacy*, 233.

37. The real value of these profits, since they were paid in Confederate currency, is probably much lower than the reported amount. See "Clover Hill Mining Company's Report to the Secretary of the Commonwealth, 23 July 1863," entry 388, Manufacturing and Mining Companies, Records of the Auditor's Office, LOV.

38. In 1863 the Niter and Mining Bureau separated from the Confederacy's War Department to constitute an independent bureau. U.S. War Department, *War of the Rebellion: A Compilation of the Official Records of the Union and Confederate Armies*, 4th ser. (Washington, D.C.: Government Printing Office, 1900), 2:594–95, 695–97; Kathleen Bruce, "Economic Factors in the Manufacture of Confederate Ordnance," *Army Ordnance* 6 (November–December 1925): 166–73, (January–February 1926): 259–64.

39. Entries of 1 March, 3 June, and 10 June 1864, Commonplace Book of Christopher Quarles Tompkins, Tompkins Family Papers, VHS; Kathleen Bruce, *Virginia Iron Manufacture in the Slave Era* (New York: Century Co., 1931), 398–99, 406–10; Eavenson, *American Coal Industry*, 444; Dew, *Ironmaker to the Confederacy*, 237; Ronald Lewis, *Coal, Iron, and Slaves: Industrial Slavery in Maryland and Virginia, 1715–1865* (Westport, Conn.: Greenwood Press, 1979), 137–39.

40. Dew, *Ironmaker to the Confederacy*, 305–7; J. B. Woodworth, "The History and Conditions of Mining in the Richmond Coal-Basin, Virginia," *Transactions of the American Institute of Mining Engineers* 31 (1902): 477–84; Oswald J. Heinrich, "The Midlothian Colliery, Virginia," *Transactions of the American Institute of Mining Engineers* 1 (May 1871–February 1873): 348; *Engineering and Mining Journal*, 1 January 1876.

41. John A. Williams, "Class, Section, and Culture in Nineteenth-Century West Virginia Politics," in *Appalachia in the Making: The Mountain South in the Nineteenth Century*, ed. Mary

Beth Pudup, Dwight B. Billings, and Altina Waller (Chapel Hill: University of North Carolina Press, 1995), 218–29; Hall, *Rending of Virginia,* 617.

42. Charles H. Ambler and Festus P. Summers, *West Virginia, the Mountain State,* 2d ed. (Englewood Cliffs, N.J.: Prentice-Hall, 1958), 142; Charleston *Western Virginian,* 25 February 1829; *Kanawha Banner* (Charleston), 24 September 1830, 15 October 1830.

43. Henry Ruffner, *Address to the People of West Virginia: Shewing That Slavery Is Injurious to the Public Welfare, and That It May Be Gradually Abolished, without Detriment to the Rights and Interests of Slaveholders* (Lexington, Va.: R. C. Noel, 1847), 12, 32, 39; William Gleason Bean, "The Ruffner Pamphlet of 1847: An Antislavery Aspect of Virginia Sectionalism," *Virginia Magazine of History and Biography* 61 (July 1953): 260–82; George Ellis Moore, "Slavery as a Factor in the Formation of West Virginia," *West Virginia History* 18 (October 1956): 56–57; William Shade, *Democratizing the Old Dominion: Virginia and the Second Party System, 1824–1861* (Charlottesville: University of Virginia Press, 1997), 209–10.

44. *Speeches of Waitman T. Willey of Monongalia County before the State Convention of Virginia, on the Basis of Representation; on County Courts and County Organization, and on the Election of Judges by the People* (Richmond: William Culley, [1851]), 15; *Anti-Slavery Reporter* (London), 1 November 1856; *Wheeling Intelligencier,* 27 March 1860; Ambler, *Sectionalism in Virginia,* 267–68; Shade, *Democratizing the Old Dominion,* 279–83.

45. Henry T. Shanks, *The Secession Movement in Virginia, 1847–1861* (1934; rpt., New York: AMS Press, 1971), 198–204.

46. Elizabeth Cometti and Festus P. Summers, *The Thirty-Fifth State: A Documentary History of West Virginia* (Morgantown: West Virginia University Library, 1966), 288–90; Shanks, *Secession Movement in Virginia,* 165–67; 204–8. Of the forty-seven delegates who represented western Virginia counties, fifteen voted for secession, twenty-eight against, and four refrained from voting. For a breakdown in the delegates and their counties, see Cometti and Summers, *Thirty-Fifth State,* 289–90.

47. Richard Curry maintained that military invasion was critical for West Virginia statehood, as only twenty-five of the fifty counties that originally made up the state supported the Confederacy. Richard Orr Curry, *A House Divided: A Study of Statehood Politics and the Copperhead Movement in West Virginia* (Pittsburgh: University of Pittsburgh Press, 1964), 52- 68.

48. Methodist circuit riders proved critical in organizing Unionist support in western Virginia during the Civil War. During the statehood movement circuit riders threw their organizational weight behind antislavery and education reform. See Sheridan Watson Bell, "The Influence of the Methodist Episcopal Church in the Establishment of the State of West Virginia" (Master's thesis, West Virginia University, 1934), 54–64.

49. The convention eventually excluded the counties in present-day southwest Virginia, on the grounds that they were "socially, economically, and traditionally inseparable from eastern Virginia." Ambler and Summers, *West Virginia,* 232.

50. Charles H. Ambler, Frances Hanoi Atwood, and William B. Mathews, eds., *Debates and*

Proceedings of the First Constitutional Convention of West Virginia (Huntington, W.Va.: Gentry Brothers, 1939), 1:438, 456, 473.

51. Ambler, Atwood, and Mathews, *Debates and Proceedings*, 2:38–39, 43, 49- 50.

52. Ambler, Atwood, and Mathews, *Debates and Proceedings*, 2:1052–53. The anti–school tax position is further developed on 1002–3, 1042–44, 1047–48, 1051–54, 1061–66, 1073–79.

53. Ambler, Atwood, and Mathews, *Debates and Proceedings*, 2:1059. For more on the pro-tax position, see 1012–14, 1045–47, 1049–50, 1056–61, 1067–69. Battelle's proposition was originally passed but later failed in committee. See 1079; Charles Ambler, *A History of Education in West Virginia from Early Colonial Times to 1949* (Huntington, W.Va.: Standard Printing and Publishing Co., 1951), 135–38 (quote is from 138).

54. Ambler, Atwood, and Mathews, *Debates and Proceedings*, 3:185. For more on the argument against extending state credit for internal improvements, see 130–33, 144–49, 151–53, 156–60, 163–66, 170–72, 175–78, 193–95, 200–202, 233–46.

55. Ambler, Atwood, and Mathews, *Debates and Proceedings*, 3:156. Also see 128–30, 153–54, 160–61, 168–70, 173–75, 189–92, 197–98, 230–33; Ambler and Summers, *West Virginia*, 235; Otis Rice, "Coal Mining in the Kanawha Valley to 1861: A View of Industrialization in the Old South," *Journal of Southern History* 31 (1965): 413–16.

56. Moore, "Slavery as a Factor in the Formation of West Virginia," 70–76. The convention's compromise slavery resolution provided for all children born of slave mothers after 1870 to be freed, the men at age twenty-eight and the women at age eighteen. It also prohibited slaves or free people of color from entering the state after the constitution was ratified. Ambler, Atwood, and Mathews, *Debates and Proceedings*, 3:371.

57. The admission of West Virginia was contingent upon the ratification of the constitution with an abolition amendment. When the constitutional convention reconvened in February 1863 to endorse the measure, pro-slavery delegates unsuccessfully attempted to get compensation for the masters of emancipated slaves. See Moore, "Slavery as a Factor in the Formation of West Virginia," 77–87.

58. *Wheeling Intelligencier*, 25 September 1863; Eavenson, *First Century and a Quarter of American Coal Industry*, 504–10; Jerry Bruce Thomas, "The Growth of the Coal Industry in the Great Kanawha Basin, 1850–1885" (Master's thesis, University of North Carolina, 1967), 50–53.

EPILOGUE

1. Waitman Willey, *Address to the People of West Virginia, on the Mineral Wealth and Agricultural Resources of the Route of the Monongahela and Lewisburg Railroad* (Washington, D.C.: n.p., 1866), 7. Pamphlet attached to the Waitman Willey Diary, WVC.

2. For an overview of American political economy for the period immediately following the Civil War, see Morton Keller, *Affairs of State: Public Life in Late Nineteenth Century America* (Cambridge: Harvard University Press, 1977), 162–96. There is a large literature regarding regulatory "capture" in the late nineteenth and twentieth centuries. For an overview, see

Thomas K. McCraw, *Prophets of Regulation: Charles Francis Adams, Louis D. Brandeis, James M. Landis, and Alfred E. Kahn* (Cambridge: Harvard University Press, 1984).

3. *United States Mining and Railroad Register*, 26 August 1865; *Miners' Journal*, 29 July 1865.

4. Job Atkins to John Wickham, 21 May 1869, Wickham Family Papers, VHS; M'Killop quoted in Howard Nicholas Eavenson, *The First Century and a Quarter of American Coal Industry* (Pittsburgh: privately printed, 1942), 130.

5. Quote is from E. J. Prescott, comp., *The Story of the Virginia Coal and Iron Company, 1882–1945* (Berryville: Virginia Book Company, [1946]), 3. For examples of promotional literature, see Hotchkiss's journal entitled *The Virginias: A Mining, Scientific, and Industrial Journal Devoted to the Development of Virginia and West Virginia* (1880–85); and his book *Virginia: A Geographical and Political Summary* (Richmond: Virginia Board of Immigration, 1876). For more on the development of southwestern Virginia, see Joseph Lambie, *From Mine to Market: The History of Coal Transportation on the Norfolk and Western Railway* (New York: New York University Press, 1954); Ron Eller, *Miners, Millhands, and Mountaineers: Industrialization of the Appalachian South, 1880–1930* (Knoxville: University of Tennessee Press, 1982), 48–52; and Kenneth Noe, *Southwest Virginia's Railroad: Modernization and the Sectional Crisis* (Urbana: University of Illinois Press, 1994).

6. B&O officials cited an act of the West Virginia legislature passed on 3 December 1863 which stated that "property exempt from taxation by the charters of any such company shall not be assessed," whereas state officials claimed that the West Virginia constitution's provision that "taxation shall be equal and uniform throughout the state" made such an exemption unconstitutional. See *Laws of West Virginia, 1863* (Wheeling: John F. McDermot, 1863), 165; *Laws of West Virginia, 1869* (Wheeling: John Frew, 1869), 113.

7. Henry Gassaway Davis (HGD) to John King, 3 August 1868; HGD to William E. Stevenson, 7 May 1869; HGD to Nathan Goff, 7 May 1869; HGD to J.W. Garrett and John King, 27 May 1869, Henry Gassaway Davis Papers, WVC. West Virginia's 1872 constitution clearly outlined the prerogative of the state to tax railroads, but both firms invoked exemptions contained in their charter. In 1882 the West Virginia State Supreme Court of Appeals ruled that these exemptions were not valid. Cometti and Summers, *Thirty-Fifth State*, 471–74.

8. Glenn Frank Massay, "Coal Consolidation: Profile of the Fairmont Field of Northern West Virginia, 1852–1903" (Ph.D. diss., West Virginia University, 1970), 9–12. The old mine operator was A. B. Fleming, who later became governor of West Virginia. See Fleming, "A History of the Fairmont Coal Region," in West Virginia Coal Mining Institute, *Proceedings, 1911* (Fairmont, W.Va.: Fairmont Printing and Lithographing Co., 1911), 251, 252.

9. In 1873 the improvement of the Kanawha was placed under federal supervision, and in 1881 all improvements on the river were surrendered to the federal government. Phil Conley, *History of the West Virginia Coal Industry* (Charleston, W.Va.: Education Foundation, 1960), 55.

10. *Laws of West Virginia, 1867* (Wheeling: John Frew, 1868), 102–5.

11. *Laws of West Virginia, 1865* (Wheeling: John Frew, 1866), 141–266; *Laws of West Vir-*

ginia, 1866 (Wheeling: John Frew, 1867), 141–266; *Laws of West Virginia, 1867* (Wheeling: John Frew, 1868), 102–5, 173–214; *Laws of West Virginia, 1868* (Wheeling: John Frew, 1869), 115–45; *Laws of West Virginia, 1869* (Wheeling: John Frew, 1870), 163–220; *Laws of West Virginia, 1872–73* (Charleston: Henry S. Walker, 1873), 775–888; *Laws of West Virginia, 1875* (Charleston: Henry S. Walker, 1875), 252–429.

12. Anthony F. C. Wallace, *St. Clair: A Nineteenth-Century Coal Town's Experience with a Disaster-Prone Industry* (New York: Alfred A. Knopf, 1987), 296–302.

13. *Laws of Pennsylvania, 1869* (Harrisburg: B. Singerly, 1869), 852–56; *Pennsylvania Archives*, 4th ser., vol. 8: *Papers of the Governors, 1858–1871* (Harrisburg: W. Stanley Ray, 1902), 1025; Alexander Trachtenberg, *The History of Legislation for the Protection of Coal Miners in Pennsylvania, 1824–1915* (New York: International Publishers, 1942), 38–39.

14. Trachtenberg, *Legislation for the Protection of Coal Miners in Pennsylvania*, 48–72.

15. See the auditor general's report for 1868, appended to *Pennsylvania Executive Documents, 1868* (Harrisburg: B. Singerly, 1869); Priscilla Long, *Where the Sun Never Shines: A History of America's Bloody Coal Industry* (New York: Paragon House, 1989), 98–99; William A. Russ Jr., "The Origin of the Ban on Special Legislation in the Constitution of 1873," *Pennsylvania History* 11 (1944): 260–75. The Coal and Iron Police's history is outlined in J. P. Shalloo, *Private Police: With Special Reference to Pennsylvania* (Philadelphia: American Academy of Political and Social Science, 1933), 58–65.

16. *Argument of Franklin B. Gowen, Esq., before the Joint Committee of the Legislature of Pennsylvania, Appointed to Inquire into the Affairs of the Philadelphia and Reading Coal and Iron Company and the Philadelphia and Reading Railroad Company, at Atlantic City, New Jersey, July 29th and 30th, 1875, on Behalf of Said Company* (Philadelphia: Press of Helfenstein, Lewis, and Greene, 1875), 72; Marvin W. Schlegel, "America's First Cartel," *Pennsylvania History* 13 (1946): 1–16; C. K. Yearley Jr., *Enterprise and Anthracite: Economics and Democracy in Schuylkill County, 1820–1875* (Baltimore: Johns Hopkins Press, 1961), 197–213.

17. Kenneth Warren, *Triumphant Capitalism: Henry Clay Frick and the Industrial Transformation of America* (Pittsburgh: University of Pittsburgh Press, 1996), 21–55.

18. Arthur A. Socolow, ed., *The State Geological Surveys: A History* (Grand Forks, N.D.: Association of American State Geologists, 1988), 376; Paul Lucier, "Commerical Interest and Scientific Disinterestedness: Consulting Geologists in Antebellum America," *Isis* 86 (1995): 245–67. Lesley distanced his survey from Rogers's in his history of the first survey entitled *A Historical Sketch of Geological Explorations in Pennsylvania and Other States* (Harrisburg: Second Geological Survey, 1876), 53–127.

19. S. Harris Daddow to Governor Boreman, 20 November 1866, *Calendar of the Arthur I. Boreman Letters in the State Department of Archives and History* (Charleston: Historical Records Survey, 1939), 58; *West Virginia Senate Journal, 1867* (Wheeling: John Frew, 1867), 16, 65; *Proceedings of the West Virginia Historical Society* (Morgantown: Morgan and Hoffman, 1871), 16; *United States Mining and Railroad Register*, 1 April 1871.

20. Maud Fulcher Callahan, "Evolution of the Constitution of West Virginia," in *West Vir-*

ginia University Studies in West Virginia History, nos. 1–2, ed. J. M. Callahan (Morgantown: West Virginia University Department of History and Political Science, 1909), 29. In exchange for a vote to overturn former Confederate proscription, Democrats withdrew opposition to Republican candidates in some instances. See Henry Gassaway Davis to George Harmond, 25 September 1869, Henry Gassaway Davis Papers, WVC.

21. David Goff to Gideon D. Camden, 22 December 1867; Samuel Price to Gideon D. Camden, 27 March 1871, Gideon D. Camden Papers, WVC; Richard O. Curry, "Crisis Politics in West Virginia, 1861–1870," in *Radicalism, Racism, and Party Realignment: The Border States during Reconstruction,* ed. Richard O. Curry (Baltimore: Johns Hopkins Press, 1969), 99–100. West Virginia's African-American population in 1870 stood at 17,980, about 4 percent of the state's total population of 442,014. U.S. Census Office, *A Compendium of the Ninth Census* (Washington, D.C.: Government Printing Office, 1872), 20.

22. *Wheeling Intelligencer,* 10 July 1871.

23. Callahan, "Evolution of the Constitution of West Virginia," 29–32; Ronald Lewis, *Transforming the Appalachian Countryside: Railroads, Deforestation, and Social Change in West Virginia, 1880–1920* (Chapel Hill: University of North Carolina Press, 1998), 104; Samuel Woods to Isabella Woods, 25 February 1872, Samuel Woods Papers, WVC. For more on the importance of the oral voting issue, see Williams, "New Dominion and the Old," 369–71.

24. *Kanawha Daily* (Charleston), 10 February 1872, 12 February 1872, 13 February 1872, 14 February 1872, 12 March 1872.

25. T. S. W. Ogden to Gideon D. Camden, 20 November 1872, 14 December 1872; Gideon D. Camden Papers, WVC; the *New York Times* quote appears in Williams, "New Dominion and the Old," 367; Lewis, *Transforming the Appalachian Countryside,* 105; Eavenson, *First Century and a Quarter of American Coal Industry,* 428.

26. Oscar Doane Lambert, *West Virginia and Its Government* (Boston: D. C. Heath, 1951), 359–61. John Alexander Williams, *West Virginia and the Captains of Industry* (Morgantown: West Virginia University Library, 1976); Eller, *Miners, Millhands, and Mountaineers,* 39–85; Lewis, *Transforming the Appalachian Countryside.*

27. Roberta Stevenson Turney, "The Encouragement of Immigration in West Virginia," *West Virginia History* 12 (October 1950): 53–55; *Fifth Annual Report of the Commissioner of Immigration of the State of West Virginia, for the Year 1868* (Wheeling: John Frew, 1868), 7. See Debar's *Fifth Annual Report,* 7; and Joseph H. Diss Debar to Gideon D. Camden, 3 January 1868, Gideon D. Camden Papers, WVC. The impact of political corruption is a major theme in the early work of Mark W. Summers; see *Railroads, Reconstruction, and the Gospel of Prosperity: Aid under the Radical Republicans, 1865–1877* (Princeton: Princeton University Press, 1984); *The Plundering Generation: Corruption and the Crisis of Union, 1849–1861* (New York: Oxford University Press, 1987); and *The Era of Good Stealings* (New York: Oxford University Press, 1993).

28. *Argument of Franklin B. Gowen, Esq., before the Joint Committee of the Legislature of Pennsylvania,* 72.

29. David Nye, *Consuming Power: A Social History of American Energies* (Cambridge, Mass.: MIT Press, 1998), 78. For more on the environmental cost of dependence upon mineral fuel in the nineteenth century, see David Stradling and Peter Thorsheim, "The Smoke of Great Cities," *Environmental History* 4 (1999): 6–31; David Stradling, *Smokestacks and Progressives: Environmentalists, Engineers, and Air Quality in America, 1881–1951* (Baltimore: Johns Hopkins University Press, 1999); Barbara Freese, *Coal: A Human History* (New York: Perseus Publishing, 2003), 71–161.

Essay ON *Sources*

SECONDARY SOURCES ON POLITICAL ECONOMY IN THE NINETEENTH CENTURY

The study of state-level political economy by no means constitutes uncharted waters for the historian, and this study builds upon a long tradition of scholarship. An exhaustive account of this literature would fill a separate volume, but several works of influence deserve special mention. A series of monographs sponsored by the Social Science Research Council (SSRC) and published in the 1940s and 1950s, for example, examined the interaction of politics and early economic development at the level of the individual state, thus giving birth to what became known as the "Commonwealth" model of antebellum political economy. The SSRC created the Committee on Research in Economic History in 1940, which sought to promote scholarship in economic history by sponsoring academic research in the role of government in antebellum economic and business development, the role of corporations and banks in American industrialization, and the changing nature of entrepreneurship in American history. See Oscar Handlin and Mary Flug Handlin, *Commonwealth: A Study of the Role of Government in the American Economy: Massachusetts, 1774–1861* (Cambridge: Harvard University Press, 1947); Louis Hartz, *Economic Policy and Democratic Thought: Pennsylvania, 1776–1860* (Cambridge: Harvard University Press, 1948); Milton Heath, *Constructive Liberalism: The Role of the State in the Economic Development of Georgia to 1860* (Cambridge: Harvard University Press, 1954); James N. Primm, *Economic Policy in the Development of a Western State: Missouri, 1820–1860* (Cambridge: Harvard University Press, 1954). For more on the Commonwealth School, see Arthur Cole, "Committee on Research in Economic History: A Description of Its Purposes, Activities, and Organization," *Journal of Economic History* 13 (1953): 79–87; and "The Committee on Research in Economic History: An Historical Sketch," *Journal of Economic History* 30 (December 1970): 723–24; Harry Scheiber, "Government and the Economy: Studies of the 'Commonwealth' Policy in Nineteenth-Century America," *Journal of Interdisciplinary History* 3 (1972): 135–51. For a critique of the SSRC works, see Robert A. Lively, "The American System: A Review Article," *Business History Review* 29 (March 1955): 81–96. Carter Goodrich and Harry Scheiber's masterful works on American canal and railroad policies solidified the Commonwealth position that state-level politics provided the framework for economic growth in the first century of American independence. See Carter Goodrich, *Government Promotion of American Canals and Railroads, 1800–1890* (New York: Columbia University Press, 1960); Harry Scheiber, *Ohio Canal Era: A Case Study of Government and the Economy, 1820–1861* (Athens: Ohio University Press, 1969).

The questions raised by the Commonwealth approach concerning the interaction of economic and political institutions, however, fell victim to the widening gap between the interests of economic and political historians. Beginning in the 1960s, the "new economic historians" explored the location and nature of industrial growth armed with new methods of quantitative analysis, which effectively removed the state-centered approach from the analysis of economic development in the antebellum era. Economic historians turned their attention to divergence but dropped political institutions from this calculus. In *The Economic Growth of the United States, 1790–1860* (New York: W. W. Norton, 1966) Douglass North argued that interregional trade between northern, southern, and western sections of the United States widened sectional divergence during the antebellum era. Peter Temin, *The Jacksonian Economy* (New York: W. W. Norton, 1969); Diane Lindstrom, *Economic Development in the Philadelphia Region, 1810–1850* (New York: Columbia University Press, 1978); and Gavin Wright, *The Political Economy of the Cotton South: Households, Markets and Wealth in the Nineteenth Century* (New York: W. W. Norton, 1978) revised North's original model to account for more localized market factors but also continued to shift the focus on divergence away from state-level institutions. Later studies of southern "distinctiveness" by economic historians virtually ignored the actions of state governments in favor of other causal factors such as natural endowments, capital flows, and culture. Northern industrialization, moreover, appeared more tied to urbanization, immigration, and entrepreneurial skill than to public policy. See, for example, Robert F. Dalzell, *Enterprising Elite: The Boston Associates and the World They Made* (Cambridge: Harvard University Press, 1987); Philip Scranton, *Proprietary Capitalism: The Textile Manufacture at Philadelphia, 1800–1885* (Cambridge: Cambridge University Press, 1983); and Sean Wilentz, *Chants Democratic: New York City and the Rise of the American Working Class, 1788–1850* (New York: Oxford University Press, 1984).

The inquiry into the compatibility of slavery and capitalism informed another strand of scholarship regarding regional economic divergence. Eugene Genovese, in his 1965 collection of revised essays entitled *The Political Economy of Slavery: Studies in the Economy and Society of the Slave South* (New York: Vintage Books, 1967), suggested that slavery perpetuated aristocratic traditions and an ideology unsuited to advanced capitalism, a charge that was repeated by Barrington Moore's magisterial work, *Social Origins of Dictatorship and Democracy: Lord and Peasant in the Making of the Modern World* (Boston: Beacon Press, 1966), 111–55; and John Ashworth,, *Slavery, Capitalism, and Politics in the Antebellum Republic*, vol. 1: *Commerce and Compromise, 1820–1850* (Cambridge: Cambridge University Press, 1995). The portrayal of acquisitive and growth-minded slave owners found in James Oakes, *The Ruling Race: A*

History of American Slaveholders (New York: Vintage Books, 1983); and *Slavery and Freedom: An Interpretation of the Slave South* (New York: Vintage Books, 1990), differed substantially from Genovese's model. Works on industrial slavery, such as Robert Starobin, *Industrial Slavery in the Old South* (New York: Oxford University Press, 1970); Ronald Lewis, *Coal, Iron, and Slaves: Industrial Slavery in Maryland and Virginia, 1715–1865* (Westport, Conn.: Greenwood Press, 1979); Charles Dew, *Bond of Iron: Master and Slave at Buffalo Forge* (New York: W. W. Norton, 1994); and John Bezís-Selfa, *Forging America: Ironworkers, Adventurers, and the Industrious Revolution* (Ithaca, N.Y.: Cornell University Press, 2004), helped recast slavery as a malleable institution capable of adapting to capitalism and enriched our understanding of economic divergence in the nineteenth century. But, like economic historians and unlike the Commonwealth historians, those who worked on the political economy of slavery rarely acknowledged the role of state government in their analysis. Because slavery was so malleable, it appeared that the presence or absence of the "peculiar institution" hardly explained regional divergence in and of itself. Studies of slavery and southern politics, moreover, tend to focus upon slavery's cultural impact and rarely delve into state-level political issues. See, for example, William Cooper, *The South and the Politics of Slavery* (Baton Rouge: Louisiana State University Press, 1978); Allen Kaufman, *Capitalism, Slavery, and Republican Values: Antebellum Political Economists, 1819–1848* (Austin: University of Texas Press, 1982); Kenneth Greenberg, *Masters and Statesmen: The Political Culture of American Slavery* (Baltimore: Johns Hopkins University Press, 1985). Roger Ransom's excellent survey of the national political economy, *Conflict and Compromise: The Political Economy of Slavery, Emancipation, and the American Civil War* (Cambridge: Cambridge University Press, 1989), is centered largely in the federal area. Two books that do analyze the impact of slavery upon state-level economic development are Alison Goodyear Freehling's examination of Virginia's bitter sectional debate over the future of slavery in the Old Dominion entitled *Drift toward Dissolution: The Virginia Slavery Debate of 1831–1832* (Baton Rouge: Louisiana State University Press, 1982); and John Majewski's excellent comparative work in economic history, *A House Dividing: Economic Development in Pennsylvania and Virginia before the Civil War* (New York: Cambridge University Press, 2000).

Political historians traditionally focus upon party formation and behavior rather than state-level policy in their depiction of nineteenth-century politics. The "new political history" of the 1970s and 1980s abandoned the field's traditional focus upon public institutions and instead examined the ethnic and cultural factors involved in party identification, electoral politics, and the role of "critical elections" in the shifting of party allegiances among the American voting population. In his book of essays entitled *The Party Period and Public Policy: American Politics from the Age of Jackson to the Progressive Era* (New York: Oxford University Press, 1986) Richard McCormick called for a change in the direction of political historians. "It is not elections (critical or otherwise) that have been the 'mainsprings' of American politics," McCormick argued; "it is government, including popular expectations for governmental actions and the rules and opportunities for getting and using the power of the State" (18). Despite excellent models for this approach, including J. Mills Thornton III, *Politics and Power in a Slave Society: Alabama, 1800–1860* (Baton Rouge: Louisiana State University Press, 1978); and L. Ray Gunn, *The Decline of Authority: Public Economic Policy and Political Development in New York State, 1800–1860* (Ithaca: Cornell University Press, 1988), few historians political heeded McCormick's recommended change of course in their examination of nineteenth-century politics. "The primacy of political parties was the dominant fact of this political era (and of no other)," Joel Silbey argued in his synthesis of American political history entitled *The American Political Nation, 1838–1893* (Stanford: Stanford University Press, 1991). "Parties defined the terms of political confrontation and shaped the behavior of most participants in the many levels of political activity" (9). Even as the field departed from the ethnocultural approach in the 1990s, the analysis of governance and political institutions remained at the margins of most political history.

While the aims and interests of political and economic historians parted from the Commonwealth approach, however, at least two groups of scholars remained focused upon the intersection of economic development and public policy. Older works in legal history such as J. Willard Hurst, *Law and the Conditions of Freedom in the Nineteenth-Century United States* (Madison: University of Wisconsin Press, 1956); and *Law and Economic Growth: The Legal His-*

tory of the Lumber Industry in Wisconsin, 1836–1915 (Cambridge: Harvard University Press, 1964); and Harry Scheiber, "Federalism and the American Economic Order, 1789–1910," *Law and Society Review* 10 (Fall 1975): 57–118; and "Regulation, Property Rights, and Definition of 'The Market': Law and the American Economy," *Journal of Economic History* 41 (March 1981): 103–9, examined antebellum America's political and legal structure's role in economic growth and revealed critical linkages between the political and economic spheres. Other studies of capitalism and the law during the nineteenth century include Morton J. Horwitz, *The Transformation of American Law, 1780–1860* (Cambridge: Harvard University Press, 1977); and Herbert Hovenkamp, *Enterprise and American Law, 1836–1937* (Cambridge: Harvard University Press, 1991). More recent works in this vein, such as Arthur McEvoy, *The Fisherman's Problem: Ecology and Law in the California Fisheries, 1850–1980* (Cambridge: Cambridge University Press, 1986); Robin L. Einhorn, *Property Rules: Political Economy in Chicago, 1833- 1872* (Chicago: University of Chicago Press, 1991); and William Novak, *The People's Welfare: Law and Regulation in Nineteenth-Century America* (Chapel Hill: University of North Carolina Press, 1996), focused upon the role of political institutions in order to understand how common law has shaped economic growth. Although they often sought answers to questions based in the American legal tradition, this line of scholarship adds to our understanding of how institutions create "pathways" for economic development to occur and provided a critical link between the Commonwealth approach of the 1940s and 1950s and more recent scholarship.

As an economic institution created by states, the business corporation is an important linkage between economic and political history. The corporation's role in American business is addressed at a general level in Glenn Porter, *The Rise of Big Business*, 2d ed. (Arlington Heights, Ill.: Harlan Davidson, 1992); Scott R. Bowman, *The Modern Corporation and American Political Thought: Law, Power, and Ideology* (University Park: Pennsylvania State University Press, 1996); William G. Roy, *Socializing Capital: The Rise of the Large Industrial Corporation in America* (Princeton: Princeton University Press, 1997). On the early history of corporations in the United States, see Pauline Maier, "The Revolutionary Origins of the American Corporation," *William and Mary Quarterly*, 3d ser., 50 (1993): 51–84; and Oscar and Mary Handlin,

"Origins of the American Business Corporation," *Journal of Economic History* 5 (1945): 1–23. A. Tony Freyer argues that the antebellum critique of corporate privileges originated in a larger struggle between "capitalist" and "producer" visions of constitutional values of legitimacy and accountability in *Producers versus Capitalists: Constitutional Conflict in Antebellum America* (Charlottesville: University of Virginia Press, 1994), 10–11. This approach can take either a celebratory or a critical angle. For the former, see Ronald Seavoy, *The Origins of the American Business Corporation, 1784–1855: Broadening the Concept of Public Service during Industrialization* (Charlottesville: University of Virginia Press, 1970). For the latter view, see Morton Horwitz, *The Transformation of American Law, 1780–1860* (Cambridge: Harvard University Press, 1977). The best works of legal history in terms of placing the legal status of the corporation within its social and political context are those of James Willard Hurst, especially *The Legitimacy of the Business Corporation in the Law of the United States, 1790–1970* (Charlottesville: University of Virginia Press, 1970); and Herbert Hovenkamp, *Enterprise and American Law, 1780–1860* (Cambridge: Harvard University Press, 1991).

The "new institutionalist" school derives theoretical inspiration from sociologist Theda Skocpol's call to "bring the state back in" to the study of politics and demonstrate a keen interest in the interaction of political and economic institutional change. Alfred Chandler, *The Visible Hand: The Managerial Revolution in American Business* (Cambridge: Harvard University Press, 1977), provides a firm-centered approach to U.S. industrial development. In his more recent work, however, he attributes differences in managerial capitalism to national institutional contexts. See his immense study entitled *Scale and Scope: The Dynamics of Industrial Capitalism* (Cambridge: Harvard University Press, 1990). Scholars of nineteenth-century development found political institutions a critical component of industrial growth. In her comparative work entitled *Politics and Industrialization: Early Railroads in the United States and Prussia* (Princeton: Princeton University Press, 1994) Colleen Dunlavy highlighted the impact of the federal state structure of the United States and suggested that important economic policy decisions in the U.S. economy occurred at the state, not the national, level. Freyer, *Producers versus Capitalists,* pursued a similar line of reasoning in his study of constitutional doctrine and economic growth in the early American republic, arguing that political institutions helped preserve "producer" interests and values in the face of an emerging American capitalism. Even Douglass North, whose early research helped trigger the quantitative revolution in economic history, stressed the importance of institutions in his more recent work. Institutions, North argued in *Institutions, Institutional Change and Economic Performance* (Cambridge: Cambridge University Press, 1990), "are the key to understanding the interrelationship between the polity and the economy and the consequences of that interrelationship for economic growth (or stagnation and decline)" (118). Richard Sylla's assertion that "government in the United States even in the nineteenth century was a growth sector" suggests that economic historians understand the vital role played by political institutions in shaping growth. See Richard Sylla, "Experimental Federalism: The Economics of American Government, 1789–1914," in *The Cambridge Economic History of the United States*, vol. 2: *The Long Nineteenth Century*, ed. Stanley L. Engerman and Robert E. Gallman (New York: Cambridge University Press, 2000): 483–541; quote is from p. 508.

Studies in American political development also reflected growing interest in institutional change. The approach of some recent works in political science to nineteenth-century state formation in the United States suggested that the study of American politics could be expanded beyond elections and voter behavior to include administrative units and their governing capacity for economic development. This trend began with Stephen Skowronek, *Building a New American State: The Expansion of National Administrative Capacities, 1877–1920* (New York: Cambridge University Press, 1982), and continued with the work of Richard Bensel, in *Yankee Leviathan: The Origins of Central State Authority in America, 1859–1877* (New York: Cambridge University Press, 1990); and *The Political Economy of American Industrialization, 1877–1900* (New York: Cambridge University Press, 2000). See also the work of economic sociologists, such as Frank Dobbin, *Forging Industrial Policy: The United States, Britain, and France in the Railway Age* (New York: Cambridge University Press, 1994); and the essays included in John L. Campbell, J. Rogers Hollingsworth, and Leon N. Lindberg, eds., *Governance of the American Economy* (New York: Cambridge University Press, 1991).

Historians of American government can draw upon a number of ground-breaking studies in nineteenth-century political economy. Drew McCoy, *The Elusive Republic: Political Economy in Jeffersonian America* (New York: W. W. Norton, 1982), provides a nuanced view of national economic policy making in the formative years of the United States as well as the major conflicts involved in the political economy of the early republic. Richard John, *Spreading the News: The American Postal System from Franklin to Morse* (Cambridge: Harvard University Press, 1995), demonstrates the critical role of federal policy in developing a national market and helping to usher in a "communications revolution." In *Internal Improvement: National Public Works and the Promise of Popular Government in the Early United States* (Chapel Hill: University of North Carolina Press, 2001) John Larson suggests that polities shaped the antebellum transportation system as much as geography. He concludes that the failures of state-level antebellum programs inevitably led to the laissez-faire fetish of later-nineteenth-century transportation policy but that both the federal and state governments struggled admirably to encourage both economic growth and republican virtue throughout this period. Finally, Heather Cox Richardson, *The Greatest Nation on Earth: Republican Economic Policies during the Civil War* (Cambridge: Harvard University Press, 1997), highlights the intersection of Republican policy making and ideology during turbulent decades of American history. These works dealing with the formation of government policy in the nineteenth century replaced the idea of a bare-boned state of "courts and parties" with a more complex vision of actors and institutions that sometimes work in tandem and sometimes work against one another. In the end a greater appreciation for the machinations of government institutions served as a welcome change from the simple instrumentalism or pluralistic models of past generations. The historiographic legacy of state-level political economy boasts a long and distinguished genealogy but will continue to offer challenges for future historians of the Industrial Revolution in the United States.

PRIMARY SOURCES IN THE POLITICAL ECONOMY OF COAL

The reconstruction of state-level political economies requires a number of disparate sources. Some of these are printed; some are held in archives; oth-

ers can be accessed on microfilm. The theme that binds them all together is
their incompleteness, as these sources rarely present a complete picture of
nineteenth-century policy making on their own. The lengthy citations through
the notes provide a comprehensive portrait of the documentation used in this
book, but I want to mention a few categories of primary sources that I found
most helpful to my research.

In a study of state-level political economy, it makes sense to begin with
state-level documents. The State Historical Society of Wisconsin's library
holds a nearly complete collection of nineteenth-century pamphlet laws, legislative records, executive and legislative documents, and constitutional convention proceedings for Virginia, West Virginia, and Pennsylvania. Since state
documents are often bound together in volumes without rhyme or reason, I
eventually developed a second sense for where important annual reports from
organizations such as the Virginia Board of Public Works or the Pennsylvania
Canal Commissioners could be found. The best way to learn these collections
is through intense browsing, but a few research tools can help give direction.
For documents from Pennsylvania's executive branch, I often consulted the
Pennsylvania Archives, 4th ser., vols. 5–8: *Papers of the Governors* (Harrisburg:
Stanley Ray, State Printer, 1900–1902). H. W. Flournoy's edited collection of
Virginia state documents, the *Calendar of Virginia State Papers and Other
Manuscripts from January 1, 1808, to December 31, 1835*, vol. 10 (Richmond:
H. W. Flournoy, 1892), helped me find some critical correspondence regard-
ing state geological surveys and other matters. The Pennsylvania and Virginia
geological surveys' annual reports were published as government documents,
and the culmination of the Pennsylvania survey can be found in Henry
Darwin Rogers, *The Geology of Pennsylvania: A Government Survey*, 2 vols.
(Philadelphia: J. B. Lippincott, 1858). I would not have been able to track chartering activity for Pennsylvania coal corporations without Calvin G. Beitel, *A
Digest of Titles of Corporations Chartered by the Legislature of Pennsylvania between the Years 1700 and 1873 Inclusive* (Philadelphia: John Campbell and Son,
1874). This index served as an aid in trolling the voluminous *Laws of Pennsyl-
vania* for both initial charters and later supplements. Finally, the story of West
Virginia's state formation can be traced in the three dense volumes of Charles
H. Ambler, Frances Hanoi Atwood, and William B. Mathews's edited collec-

tion entitled *Debates and Proceedings of the First Constitutional Convention of West Virginia* (Huntington, W.Va.: Gentry Brothers, 1939).

State documents help establish the general parameters of state-level political economies of coal, but they do not reveal much about the politicians and colliers who helped breathe life into these policies. Although I cannot list the numerous manuscript collections that allowed me to add substance to the bare skeletal outlines provided by government sources, a few important repositories and collections deserve mention here. The University of Virginia's Alderman Library holds the single most important manuscript collection for the early Virginia coal industry, the Henry Heth Papers. Several important Heth letters can be found at the Virginia Historical Society as well as those of another important coal mining family in the Tompkins Family Papers. The Library of Virginia holds several important manuscript collections for the Virginia coal industry; the most important of these that I consulted was the rather large Legislative Petitions File and the Board of Public Works Collection. The American Philosophical Society's J. P. Lesley Papers and William B. Foulkes Papers contain important letters from individual colliers as well as records pertaining to the geological surveys of Pennsylvania. The Slifer-Dill Collection and the Fall Brook Coal Company Records, held in the Pennsylvania State Archives in Harrisburg, provided a variety of perspectives on the coal trade. Another invaluable source in Harrisburg is the collection of corporate charters found in the Records of the Department of State, Corporations Bureau, Letters Patent, 1814–74. The West Virginia Collection in Morgantown holds many collections pertinent to that state's foundation and early coal industry. Among the most important that I consulted were the Henry Gassaway Davis Papers and the Waitman Willey Papers.

Several printed primary sources aided in the reconstruction of the nineteenth-century political economy of coal, although they tend to appear later in time. Some of the most important include Richard C. Taylor, *Statistics of Coal: The Geographical and Geological Distribution of Mineral Combustibles or Fossil Fuel* (Philadelphia: J. W. Moore, 1848); George H. Thurston, *Directory of the Monongahela and Youghiogheny Valleys: Containing Brief Historical Sketches of the Various Towns Located on Them; with a Statistical Exhibit of the Collieries upon the Two Rivers* (Pittsburgh: A. A. Anderson, 1859); Samuel Harries Daddow, *Coal, Iron, and Oil, or the Practical American Miner* (Pottsville, Pa.: Benjamin Bannan, 1866); and James McFarlane, *The Coal Regions of America* (New York: D. Appleton, 1873). The Hagley Museum and Library in Greenville, Delaware, holds a wealth of printed coal and railroad company prospectuses, pamphlets by Henry Carey and other noted political economists, and a rich collection of pamphlets regarding the early conflict between the Lehigh Coal and Navigation Company and Schuylkill County interests. Although much of the material on geological surveys exists in manuscript form, William B. Rogers's widow, Emma Rogers, completed the work of her late husband by printing the annual reports of his survey in his volume entitled, *A Reprint of Annual Reports and Other Papers, on the Geology of the Virginias* (New York: D. Appleton, 1884); as well as his correspondence in her edited two-volume collection entitled *Life and Letters of William Barton Rogers* (Boston: Houghton Mifflin, 1896).

Of course, no discussion of the nineteenth-century coal trade would be complete without mention of the important work of Howard Eavenson. The former mining engineer devoted years to the compilation and publication of materials regarding the American coal industry. His articles and books on specific coal regions serve more as collections of primary documents than analytical scholarship, but they remain invaluable sources to historians of coal. They include *The Pittsburgh Coal Bed: Its Early History and Development.* (New York: American Institute of Mining and Metallurgical Engineers, 1938); "Some Side-Lights on Early Virginia Coal Mining," *Virginia Magazine of History and Biography* 5 (1942): 199–208; and "Notes on an Old West Virginia Coal Field," *West Virginia History* 5 (1944): 83–100. His magnum opus is a comprehensive look at American coal entitled *The First Century and a Quarter of the American Coal Industry* (Pittsburgh: privately printed, 1942). This work mostly contains reprints of letters, newspaper articles, and excerpts from books regarding the coal industry of each state, but Eavenson also attempted to compile the most comprehensive set of production statistics. Measuring coal production in the nineteenth century is more an art than science, as units vary from bushels, chaldrons, and short and long tons of varying size. I share his skeptical outlook on the finality of these measures and tried to shore up the limited quantitative analysis in this book with corroborative accounts from qualitative sources. But, with this caveat in mind, I found Eavenson's estimates priceless, and all industrial historians owe him a great debt.

Whenever I needed to flesh out a story or controversy, I relied upon the various newspapers and journals that dealt with the coal trade during this period. Newspapers of various sizes and stripes litter the footnotes of this book, but a few deserve special mention. Benjamin Bannan's *Miners' Journal*, published in Pottsville, Pennsylvania, was the leading coal trade journal of the antebellum period. Philadelphia's *United States Rail Road and Mining Register* emerged as an important journal in the 1850s. Periodicals with a national focus which were most helpful include the *American Journal of Science* (Philadelphia), *Hunt's Merchant Magazine* (New York), and *Niles' Register* (Baltimore). When consulted in tandem, these contemporary accounts provided important statistical, political, and social commentary on the coal trade.

Index

Company, 100–101, 187; Pennsylvania
Coal Company, 197–98; Pittsburgh and
Youghiogheny Coal Company, 202;
Powhatan Coal Company, 167; Preston
Railroad, Lumber, and Mining Company
of Preston County, 179; Raymond Mining
Company, 120; Saint George Mining and
Manufacturing Company, 185; Towanda
Coal Company, 201; Virginia Coal and
Iron Company, 226–27; West Virginia
Coal Mining Company, 179; Wolf Creek
Diamond Coal Company, 200

coalfields: Broad Top bituminous, 51, 139,
149; Connellsville coke region, 80, 232–33;
Lehigh anthracite region, 53, 139; Richmond bituminous, 1–2, 25–27, 82–83,
208–10, 225–26; Pennsylvania anthracite,
2–3, 52–53; Phillipsburg semi-bituminous
region, 201; Pittsburgh Seam bituminous,
51–52, 81–82, 134, 149, 202; Schuylkill anthracite region, 133–34, 53; western Virginia bituminous, 3, 27–29, 86–89, 178–
79, 186–87; West Virginia bituminous,
220–21, 237; Wyoming or northern anthracite region, 139, 175

coke. See coal, bituminous

Confederacy, 208; Niter and Mining Bureau,
209

Conscription Act (1863), 207

constitutional conventions, Pennsylvania:
1776, 62; 1790, 63–64; 1837, 105–6, 164–
68; 1873, 231–32

constitutional conventions, Virginia: 1776,
30–31; 1829–30, 90–91; 1850–51, 91, 100,
102, 180–81, 213

constitutional conventions, West Virginia:
1861, 214–221; 1872, 236–37

corporations: antebellum development,
155–57; circuit courts create in Virginia,
182–87; creation as industrial policy, 176–

78; during Civil War, 195–204; general incorporation, 171–74, 175–78; ideological
opposition to, 152–54; impact on Richmond basin, 166–68; improvement companies, 176; legal status in Virginia, 168–
69; Pennsylvania Senate investigates role
in coal trade in 1834, 161–62; regulation of
"foreign" corporations, 157–58; special
incorporation, 162–64, 169; used to promote anthracite iron, 78; West Virginia
and, 229–31

Coxe, Tench, 2, 25

Curtain, Andrew, 192–93, 194, 198–99, 203,
207–8

D&H. See Delaware and Hudson Canal

Daddow, Samuel Harris, 234

Daily Dispatch (Richmond), 180

Darby, Abraham, 18

Dartmouth College v. Woodward (1819), 168

Davidson, John, 36–37

Davis, Henry Gassaway, 227

Davis, John A. G., 147

Debar, Joseph H. Diss, 238–39

Delaware and Hudson Canal Company
(D&H), 175, 202; coal traffic on, 67, 69;
competition with North Branch division
canal, 113–14; creation of, 69–70; strike
by workers, 205

Delaware division canal. See Pennsylvania
State Works

Dering, Henry, 219–20

Duncan, Stephen, 158, 160–61

DuVal, Samuel, 27

Engineering and Mining Journal, 210

Erie Extension Canal. See Pennsylvania
State Works

Escheat Act (1833), 158

Evans, Cadwalader, 66

9 780801 894008